INTERNATIONAL CENTRE FOR MECHANICAL SCIENCES

COURSES AND LECTURES - No. 282

CONTINUUM THEORY OF THE MECHANICS OF FIBRE-REINFORCED COMPOSITES

EDITED BY

A.J.M. SPENCER

THE UNIVERSITY OF NOTTINGHAM

SPRINGER-VERLAG WIEN GMBH

Originally published by Springer-Verlag Wien New York in 1984

ISBN 978-3-211-81842-8 ISBN 978-3-7091-4336-0 (eBook)
DOI 10.1007/978-3-7091-4336-0

PREFACE

This book is made up of the notes for the lectures the five authors gave at the International Centre for Mechanical Sciences in Udine in July 1981. The aim of the School was to present an account of recent developments in the mechanics of fibre-reinforced composites and other highly anisotropic materials, from the continuum mechanics point of view.

The use of fibre-reinforced composite materials in engineering applications is now well-established. In order that these materials may be used effectively, it is important to have suitable mathematical descriptions of their mechanical behaviour, and efficient methods of analysis of this behaviour. At a more fundamental level, the mechanics of highly anisotropic materials raises many interesting questions in theoretical solid mechanics which have led to important improvements in our understanding of the subject. In consequence, the lecture notes are concerned both with the formulation of equations to describe the mechanical behaviour of fibre-reinforced materials having various kinds of mechanical response, and with the application of these equations to the solution of problems of practical and theoretical interest.

I am very pleased to express the thanks of the authors to Mrs. Judith Hind, who has carried out much the greater part of the arduous task of typing and assembling this text. I am also pleased to thank the officers and the secretariat of CISM for their invitation to give the lectures, and their help in the organisation of the School.

Nottingham, August 1984 — *A. J. M. Spencer*

CONTENTS

I

CONSTITUTIVE THEORY FOR STRONGLY ANISOTROPIC SOLIDS

A. J. M. SPENCER

Department of Theoretical Mechanics
University of Nottingham
Nottingham, NG7 2RD
England

1. INTRODUCTION

We shall discuss a number of problems concerned with stress and deformation analysis of fibre-reinforced composite, and other strongly anisotropic, materials. The kind of composite material in mind is one in which a matrix material is reinforced by strong stiff fibres which are systematically arranged in the matrix. The fibres are considered to be long compared to their diameters and the fibre spacings, and to be quite densely distributed, so that the fibres form a substantial proportion (typically about 50% by volume) of the composite. There are many such composite materials now in use or under development; examples are carbon fibre reinforced epoxy resins, boron fibre reinforced aluminium, and nylon or steel reinforced rubber which is used in tyres, hosepipes and belts.

Since we assume the fibres to be systematically arranged, a composite of this kind has strong directional properties, so that macroscopically it has to be regarded as an anisotropic material. In most cases this anisotropy is very strong, so that mechanical properties are highly dependent on direction in the material; some examples of this will be given later. If the material is reinforced with a single family of fibres, which are randomly distributed in the cross sections normal to the fibres, then the composite material has a single preferred direction (which we shall call the *fibre direction*) and so is transversely isotropic with respect to this direction. The fibre direction may be characterized by a unit vector $\underset{\sim}{a}$. However, it is not necessary that the fibre direction be the same at each point; it is quite possible to align the fibres along a family of curves. Then the composite is *locally* transversely anisotropic with respect to the *local* fibre direction, and $\underset{\sim}{a}$ is a function of position.

It is also possible to have reinforcement by more than one family of fibres. For example, we may consider a laminated plate built up from a large number of thin laminae, each of which is unidirectionally reinforced, but which are stacked alternately with the fibres aligned in two different directions. On the macroscopic scale such a laminate will have two preferred directions, and so will have orthotropic symmetry. It is easy to envisage laminates with three or more preferred directions. Another configuration of interest is that of a circular cylinder reinforced by helical fibres lying in concentric circular cylindrical surfaces and wound symmetrically in opposing directions. This material is locally orthotropic but the preferred directions vary with position. In these cases we have two or more fibre directions, each of which may be characterized by a unit vector $\underset{\sim}{a}$, $\underset{\sim}{b}$, ...

We consider that the fibres are distributed throughout the material (the possibility of variations in fibre density is not excluded). There are also interesting problems in which the fibres lie in discrete surfaces, but these will not be treated here. We are concerned with the development of continuum theories, so we assume the fibres to be continuously distributed through the material. Then the fibre directions $\underset{\sim}{a}$, $\underset{\sim}{b}$, ... may be regarded as continuous functions of position $\underset{\sim}{x}$. Thus on the macroscopic scale we treat the material as a strongly anisotropic continuum. The theories and solutions we shall develop may be applied not only to fibre-reinforced materials, but also to any strongly anisotropic material. However, it is convenient to use terminology associated with fibre-reinforced materials, and so we shall refer to the directions characterizing the strong anisotropy as fibre directions, and to their trajectories as fibres.

This continuum approach excludes any consideration of the micromechanics of the composite, which involves interactions between individual fibres and the matrix. There are many important problems on the microscopic scale, but these will not be considered. Another important problem area is that of the relations between the mechanical properties of the composite and those of the constituents which form the composite. Such problems will also not be considered.

Clearly a fibre-reinforced composite material may show all kinds of mechanical response. We shall deal mainly with elastic and plastic behaviour, for both large and small deformations. There is also a substantial body of theory which does not depend on material response, aspects of which are dealt with in Chapter II.

2. LINEAR ELASTIC CONSTITUTIVE EQUATIONS FOR FIBRE-REINFORCED MATERIAL

2.1 Linear elasticity - one family of fibres

We begin with the simplest case, which is that of a linearly elastic solid reinforced by a single family of fibres. The constitutive equation is therefore that of a transversely isotropic linearly elastic solid. This constitutive equation is well known; the usual method of deriving it is to select a coordinate system such that one of the coordinate axes coincides with the axis of transverse isotropy, and examine the restrictions on the strain-energy function which result from the requirements of invariance under rotations about this axis. We proceed in a rather different, though equivalent, way. The main reason for this is that, because the fibre direction is dependent on position, it is convenient to have a formulation which does not depend on a particular choice of coordinate system. It is quite possible to transform the standard results so that they do not depend on the choice of coordinate system, but it seems more satisfactory to adopt a formulation which is coordinate-free from the outset. This is especially the case when we proceed to consider more complicated constitutive equations involving finite deformations and two families of fibres.

All vector and tensor components will be referred to a system of rectangular cartesian coordinates x_i $(i = 1,2,3)$. Components of the infinitesimal displacement vector $\underset{\sim}{u}$ are denoted by u_i, and components of the infinitesimal strain tensor $\underset{\sim}{e}$ by e_{ij}, so that

$$e_{ij} = \frac{1}{2}\left(\frac{\partial u_i}{\partial x_j} + \frac{\partial u_j}{\partial x_i}\right). \tag{1}$$

The Cauchy stress tensor $\underset{\sim}{\sigma}$ has components σ_{ij}, and the fibre direction vector $\underset{\sim}{a}$ has components a_i.

In linear elasticity, the strain-energy function W is a quadratic function of e_{ij}, so that

$$W = \tfrac{1}{2} c_{ijk\ell} e_{ij} e_{k\ell}, \tag{2}$$

where the usual repeated index summation convention is used, and $c_{ijk\ell}$ are components of the stiffness tensor, which possesses the symmetries

$$c_{ijk\ell} = c_{k\ell ij} = c_{jik\ell} = c_{ij\ell k}. \tag{3}$$

The stress is then given by

$$\sigma_{ij} = \frac{\partial W}{\partial e_{ij}} = c_{ijk\ell} e_{k\ell}. \tag{4}$$

The stiffness components $c_{ijk\ell}$ depend on the fibre direction, and so may vary with position.

To determine the form of the $c_{ijk\ell}$ for a transversely isotropic material we first note that, for a given deformation, W depends on $\underset{\sim}{e}$ and on the fibre direction $\underset{\sim}{a}$. Thus

$$W = W(\underset{\sim}{e}, \underset{\sim}{a}).$$

If the only anisotropic properties of the material are those which arise from the presence of the fibres, then W is unchanged if both the deformation field and the fibres undergo a rotation which is described by a proper orthogonal tensor $\underset{\sim}{Q}$. For this new deformation, the strain is given by $\bar{\underset{\sim}{e}} = \underset{\sim}{Q}\underset{\sim}{e}\underset{\sim}{Q}^T$, and the fibre direction by $\bar{\underset{\sim}{a}} = \underset{\sim}{Q}\underset{\sim}{a}$. Thus

$$W(\underset{\sim}{e},\underset{\sim}{a}) = W(\underset{\sim}{Q}\underset{\sim}{e}\underset{\sim}{Q}^T,\underset{\sim}{Q}\underset{\sim}{a}) \tag{5}$$

and this holds for all proper orthogonal tensors $\underset{\sim}{Q}$, that is for all tensors such that $\underset{\sim}{Q}\underset{\sim}{Q}^T = \underset{\sim}{Q}^T\underset{\sim}{Q} = \underset{\sim}{I}$, $\det \underset{\sim}{Q} = 1$, where $\underset{\sim}{I}$ is the unit tensor.

Equation (5) is a statement that W is an isotropic invariant of $\underset{\sim}{e}$ and $\underset{\sim}{a}$. Since the sense of $\underset{\sim}{a}$ is not significant, W must be an even function of $\underset{\sim}{a}$, and so it may be expressed as an isotropic invariant of $\underset{\sim}{e}$ and $\underset{\sim}{a}\otimes\underset{\sim}{a}$, where the dyadic product $\underset{\sim}{a}\otimes\underset{\sim}{a}$ is the second-order tensor with cartesian components $a_i a_j$. These invariants are tabulated (see, for example, [1]); by reading off from tables we find that W can be expressed as a function of the traces of the following tensor products:

$$\underset{\sim}{e}, \quad \underset{\sim}{e}^2, \quad \underset{\sim}{e}^3, \quad \underset{\sim}{a}\otimes\underset{\sim}{a}, \quad (\underset{\sim}{a}\otimes\underset{\sim}{a})^2, \quad (\underset{\sim}{a}\otimes\underset{\sim}{a})^3,$$

$$\underset{\sim}{e}.\underset{\sim}{a}\otimes\underset{\sim}{a}, \quad \underset{\sim}{e}.(\underset{\sim}{a}\otimes\underset{\sim}{a})^2, \quad \underset{\sim}{e}^2.\underset{\sim}{a}\otimes\underset{\sim}{a}, \quad \underset{\sim}{e}^2.(\underset{\sim}{a}\otimes\underset{\sim}{a})^2.$$

However, since $\underset{\sim}{a}$ is a unit vector

$$a\otimes a = (a\otimes a)^2 = (a\otimes a)^3 = \ldots . \tag{6}$$

Also

$$\mathrm{tr}\,\underset{\sim}{a}\otimes\underset{\sim}{a} = 1, \qquad \mathrm{tr}\,\underset{\sim}{e}.\underset{\sim}{a}\otimes\underset{\sim}{a} = \underset{\sim}{a}.\underset{\sim}{e}.\underset{\sim}{a}, \qquad \mathrm{tr}\,\underset{\sim}{e}^2.\underset{\sim}{a}\otimes\underset{\sim}{a} = \underset{\sim}{a}.\underset{\sim}{e}^2.\underset{\sim}{a} \tag{7}$$

and so the set of invariants reduces to

$$\mathrm{tr}\,\underset{\sim}{e}, \quad \mathrm{tr}\,\underset{\sim}{e}^2, \quad \mathrm{tr}\,\underset{\sim}{e}^3, \quad \underset{\sim}{a}.\underset{\sim}{e}.\underset{\sim}{a}, \quad \underset{\sim}{a}.\underset{\sim}{e}^2.\underset{\sim}{a}\,. \tag{8}$$

The most general quadratic function in $\underset{\sim}{e}$ which can be formed from (8) is

$$\begin{aligned} W &= \tfrac{1}{2}\lambda(\mathrm{tr}\,\underset{\sim}{e})^2 + \mu_T\,\mathrm{tr}\,\underset{\sim}{e}^2 + \alpha(\underset{\sim}{a}.\underset{\sim}{e}.\underset{\sim}{a})\,\mathrm{tr}\,\underset{\sim}{e} \\ &\qquad + 2(\mu_L-\mu_T)\,\underset{\sim}{a}.\underset{\sim}{e}^2.\underset{\sim}{a} + \tfrac{1}{2}\beta(\underset{\sim}{a}.\underset{\sim}{e}.\underset{\sim}{a})^2 \\ &= \tfrac{1}{2}\lambda e_{ii}e_{kk} + \mu_T e_{ik}e_{ik} + \alpha a_i e_{ij} a_j e_{kk} \\ &\qquad + 2(\mu_L-\mu_T)a_i e_{ij} e_{jk} a_k + \tfrac{1}{2}\beta a_i a_j e_{ij} a_k a_\ell e_{k\ell}\,, \end{aligned} \tag{9}$$

where λ, μ_T, μ_L, α and β are elastic constants. Thus for this case

$$\sigma_{ij} = \lambda e_{kk}\delta_{ij} + 2\mu_T e_{ij} + \alpha(a_k a_\ell e_{k\ell}\delta_{ij} + a_i a_j e_{kk})$$
$$+ 2(\mu_L - \mu_T)(a_i a_k e_{kj} + a_j a_k e_{ki}) + \beta a_i a_j a_k a_\ell e_{k\ell}, \tag{10}$$

where δ_{ij} denotes the Kronecker delta, so that

$$c_{ijk\ell} = \lambda\delta_{ij}\delta_{k\ell} + \mu_T(\delta_{ik}\delta_{j\ell} + \delta_{jk}\delta_{i\ell}) + \alpha(a_k a_\ell \delta_{ij} + a_i a_j \delta_{k\ell})$$
$$+ (\mu_L - \mu_T)(a_i a_k \delta_{j\ell} + a_i a_\ell \delta_{jk} + a_j a_k \delta_{i\ell} + a_j a_\ell \delta_{ik}) + \beta a_i a_j a_k a_\ell . \tag{11}$$

An alternative derivation of this result is given in [2]. In direct notation, (10) can be written as

$$\underset{\sim}{\sigma} = (\lambda \operatorname{tr} \underset{\sim}{e} + \alpha \underset{\sim}{a}.\underset{\sim}{e}.\underset{\sim}{a}) \underset{\sim}{I} + 2\mu_T \underset{\sim}{e} + (\alpha \operatorname{tr} \underset{\sim}{e} + \beta \underset{\sim}{a}.\underset{\sim}{e}.\underset{\sim}{a}) \underset{\sim}{a} \otimes \underset{\sim}{a}$$
$$+ 2(\mu_L - \mu_T)(\underset{\sim}{a} \otimes \underset{\sim}{a}.\underset{\sim}{e} + \underset{\sim}{e}.\underset{\sim}{a} \otimes \underset{\sim}{a}) . \tag{12}$$

The elastic constants μ_L and μ_T represent shear moduli. The other elastic constants λ, α and β can be related to other elastic constants which have more direct physical interpretations, such as extension moduli and Poisson's ratios. The admissible values of the elastic constants are restricted by the requirement that W must be positive definite.

Suppose, for example, that the direction of the x_1-axis is chosen to coincide with the fibre direction, so that $\underset{\sim}{a}$ has components (1,0,0). Then (10) give

$$\begin{pmatrix} \sigma_{11} \\ \sigma_{22} \\ \sigma_{33} \\ \sigma_{23} \\ \sigma_{31} \\ \sigma_{12} \end{pmatrix} = \begin{pmatrix} \lambda+2\alpha+4\mu_L-2\mu_T+\beta & \lambda+\alpha & \lambda+\alpha & 0 & 0 & 0 \\ \lambda+\alpha & \lambda+2\mu_T & \lambda & 0 & 0 & 0 \\ \lambda+\alpha & \lambda & \lambda+2\mu_T & 0 & 0 & 0 \\ 0 & 0 & 0 & 2\mu_T & 0 & 0 \\ 0 & 0 & 0 & 0 & 2\mu_L & 0 \\ 0 & 0 & 0 & 0 & 0 & 2\mu_L \end{pmatrix} \begin{pmatrix} e_{11} \\ e_{22} \\ e_{33} \\ e_{23} \\ e_{31} \\ e_{12} \end{pmatrix} \tag{13}$$

which is equivalent to the usual form of the constitutive equation for a linearly elastic material which is transversely isotropic with respect to the x_1-axis. From (13) it is apparent that μ_L and μ_T are shear moduli for shear on planes parallel to the fibres, with direction of shear in the fibre direction (μ_L) and normal to the fibre direction (μ_T) respectively. Also from (13) it is easily shown that the extension moduli E_L for uniaxial tension in the fibre direction and E_T for uniaxial tension in directions normal to the fibres are

$$E_L = \frac{(\lambda+\mu_T)(\beta+2\mu_L)+\mu_T(\lambda+2\alpha)-\alpha^2}{\lambda+\mu_T},$$

$$E_T = \frac{4\mu_T\{(\lambda+\mu_T)(\beta+2\mu_L)+\mu_T(\lambda+2\alpha)-\alpha^2\}}{(\lambda+2\mu_T)(\beta+2\mu_L)+2\mu_T(\lambda+2\alpha)-\alpha^2}. \qquad (14)$$

Expressions for the Poisson's ratios associated with extension in the fibre and transverse directions are also readily derived from (13).

Values of the elastic constants μ_T, μ_L, λ, α and β can be determined from experimental measurements. For example, Markham [3] obtained data for a typical carbon fibre - epoxy resin composite which give (in units of 10^9 Nm^{-2})

$$\mu_L = 5.66, \quad \mu_T = 2.46, \quad \lambda = 5.64, \quad \alpha = -1.27, \quad \beta = 227.29 \qquad (15)$$

and hence, in the same units,

$$E_L = 239.35, \quad E_T = 7.53 \,. \qquad (16)$$

Kinematic constraints. We see from (15) and (16) that the modulus E_L considerably exceeds the other extension and shear moduli. This of course reflects the stiffness of the material in the fibre direction, and is a feature of many fibre-reinforced composites. The material is resistant to deformation by extension in the fibre direction, and will prefer other deformation mechanisms if any are available it it. This suggests that as a first approximation we might consider the limit in which $E_L \to \infty$, while E_T, μ_L and μ_T remain finite. This corresponds to the case in which the material is incapable of extension in the fibre direction, so that in any deformation the strain component $a_i a_j e_{ij}$ is zero.

The condition

$$a_i a_j e_{ij} = 0 \qquad (17)$$

is an example of a *kinematic constraint*, and represents *inextensibility* in the fibre direction $\underset{\sim}{a}$. Kinematic constraints are not uncommon in continuum mechanics. The constraint which we encounter most often is that of *incompressibility*, which is often used in, for example, fluid mechanics and finite elasticity theory. Although no material is truly incompressible, there are many materials and applications for which the assumption of incompressibility gives satisfactory results. Because it is more familiar, we shall consider first the effect of considering the material to be incompressible, and return to the inextensibility constraint later.

Incompressible material. If $\operatorname{tr} \underset{\sim}{e} = 0$, then (for the transversely isotropic linearly elastic material under consideration) from (8) W becomes a function of $\operatorname{tr} \underset{\sim}{e}^3$, $\operatorname{tr} \underset{\sim}{e}^2$, $\underset{\sim}{a}.\underset{\sim}{e}.\underset{\sim}{a}$, $\underset{\sim}{a}.\underset{\sim}{e}^2.\underset{\sim}{a}$, but we may add to W any multiple of $\operatorname{tr} \underset{\sim}{e}$. Hence (9) is replaced by

$$W = \mu_T \operatorname{tr} \underset{\sim}{e}^2 + 2(\mu_L - \mu_T)\underset{\sim}{a}.\underset{\sim}{e}^2.\underset{\sim}{a} + \tfrac{1}{2}\beta(\underset{\sim}{a}.\underset{\sim}{e}.\underset{\sim}{a})^2 - p \operatorname{tr} \underset{\sim}{e} , \qquad (18)$$

where p may be regarded as a Lagrangian multiplier. The number of independent elastic constants is reduced to three. Then (4) gives (in direct notation)

$$\underset{\sim}{\sigma} = -p\underset{\sim}{I} + 2\mu_T \underset{\sim}{e} + \beta(\underset{\sim}{a}.\underset{\sim}{e}.\underset{\sim}{a})\underset{\sim}{a}\otimes\underset{\sim}{a} + 2(\mu_L - \mu_T)(\underset{\sim}{a}\otimes\underset{\sim}{a}.\underset{\sim}{e} + \underset{\sim}{e}.\underset{\sim}{a}\otimes\underset{\sim}{a}) . \qquad (19)$$

Here p is arbitrary (in the sense that it is not given by a constitutive equation but has to be determined by equations of equilibrium or motion and boundary conditions) and represents an arbitrary hydrostatic pressure. This hydrostatic pressure is a *reaction* to the constraint of incompressibility. The stress $-p\underset{\sim}{I}$ does no work in any deformation which conforms to the constraint of incompressibility, for if $e_{kk} = 0$, then

$$-p\delta_{ij} e_{ij} = -pe_{kk} = 0 .$$

From (19) we see that we may divide the stress into two parts

$$\underset{\sim}{\sigma} = \underset{\sim}{s} + \underset{\sim}{r} , \qquad \text{or} \qquad \sigma_{ij} = s_{ij} + r_{ij} , \qquad (20)$$

where $\underset{\sim}{r}$ represents the *reaction stress*, and is here of the form $-p\underset{\sim}{I}$, and $\underset{\sim}{s}$ is called the *extra-stress*. For a material subject to kinematic constraints, the extra-stress is given by constitutive equations, and the reaction stress is arbitrary in the sense described above. Since p is arbitrary, $\underset{\sim}{s}$ is arbitrary to within a hydrostatic pressure, so without loss of generality we may specify that $\operatorname{tr} \underset{\sim}{s} = 0$. Then $\underset{\sim}{s}$ becomes the deviatoric stress

$$\underset{\sim}{s} = \underset{\sim}{\sigma} - \tfrac{1}{3}\underset{\sim}{I}\,\mathrm{tr}\,\underset{\sim}{\sigma} \tag{21}$$

and from (19), and using $\mathrm{tr}\,\underset{\sim}{e} = 0$ and $\underset{\sim}{a}.\underset{\sim}{a} = 1$, we have

$$\underset{\sim}{s} = 2\mu_T\underset{\sim}{e} + \beta(\underset{\sim}{a}.\underset{\sim}{e}.\underset{\sim}{a})\underset{\sim}{a}\otimes\underset{\sim}{a} + 2(\mu_L-\mu_T)(\underset{\sim}{a}\otimes\underset{\sim}{a}.\underset{\sim}{e}+\underset{\sim}{e}.\underset{\sim}{a}\otimes\underset{\sim}{a})$$

$$- \tfrac{1}{3}(\beta+4\mu_L-4\mu_T)(\underset{\sim}{a}.\underset{\sim}{e}.\underset{\sim}{a})\underset{\sim}{I}\,. \tag{22}$$

Inextensible material. A similar procedure can be followed if the material is inextensible in the fibre direction but not incompressible. Then $\underset{\sim}{a}.\underset{\sim}{e}.\underset{\sim}{a} = 0$, and (9) is replaced by

$$W = \tfrac{1}{2}\lambda(\mathrm{tr}\,\underset{\sim}{e})^2 + \mu_T\,\mathrm{tr}\,\underset{\sim}{e}^2 + 2(\mu_L-\mu_T)\underset{\sim}{a}.\underset{\sim}{e}^2.\underset{\sim}{a} + T\underset{\sim}{a}.\underset{\sim}{e}.\underset{\sim}{a}\,, \tag{23}$$

where T is a Lagrangian multiplier, and again the number of independent elastic constants is reduced to three. Then from (4)

$$\underset{\sim}{\sigma} = T\underset{\sim}{a}\otimes\underset{\sim}{a} + \lambda\underset{\sim}{I}\,\mathrm{tr}\,\underset{\sim}{e} + 2\mu_T\underset{\sim}{e} + 2(\mu_L-\mu_T)(\underset{\sim}{a}\otimes\underset{\sim}{a}.\underset{\sim}{e}+\underset{\sim}{e}.\underset{\sim}{a}\otimes\underset{\sim}{a})\,. \tag{24}$$

The stress $T\underset{\sim}{a}\otimes\underset{\sim}{a}$ is an arbitrary tension in the fibre direction which is a reaction to the inextensibility constraint and does no work in any deformation which conforms to this constraint, for

$$Ta_ia_je_{ij} = T\underset{\sim}{a}.\underset{\sim}{e}.\underset{\sim}{a} = 0\,.$$

If we decompose $\underset{\sim}{\sigma}$ into a reaction stress $\underset{\sim}{r}$ and an extra-stress $\underset{\sim}{s}$, as in (20), then

$$\underset{\sim}{r} = T\underset{\sim}{a}\otimes\underset{\sim}{a} \tag{25}$$

and $\underset{\sim}{s}$ is arbitrary to within a fibre tension. Without loss of generality we may specify $\underset{\sim}{a}.\underset{\sim}{s}.\underset{\sim}{a} = 0$, and then

$$\underset{\sim}{s} = \lambda(\underset{\sim}{I}-\underset{\sim}{a}\otimes\underset{\sim}{a})\,\mathrm{tr}\,\underset{\sim}{e} + 2\mu_T\underset{\sim}{e} + 2(\mu_L-\mu_T)(\underset{\sim}{a}\otimes\underset{\sim}{a}.\underset{\sim}{e}+\underset{\sim}{e}.\underset{\sim}{a}\otimes\underset{\sim}{a})\,. \tag{26}$$

Incompressible and inextensible material. If the material is both incompressible and inextensible in the fibre direction, then W takes the form

$$W = \mu_T\,\mathrm{tr}\,\underset{\sim}{e}^2 + 2(\mu_L-\mu_T)\underset{\sim}{a}.\underset{\sim}{e}^2.\underset{\sim}{a} - p\,\mathrm{tr}\,\underset{\sim}{e} + T\underset{\sim}{a}.\underset{\sim}{e}.\underset{\sim}{a}\,, \tag{27}$$

and hence

$$\underset{\sim}{\sigma} = 2\mu_T\underset{\sim}{e} + 2(\mu_L-\mu_T)(\underset{\sim}{a}\otimes\underset{\sim}{a}.\underset{\sim}{e}+\underset{\sim}{e}.\underset{\sim}{a}\otimes\underset{\sim}{a}) - p\underset{\sim}{I} + T\underset{\sim}{a}\otimes\underset{\sim}{a} \ . \tag{28}$$

There are now only two independent elastic constants μ_T and μ_L. The reaction stress is

$$\underset{\sim}{r} = -p\underset{\sim}{I} + T\underset{\sim}{a}\otimes\underset{\sim}{a} \ , \tag{29}$$

and $\underset{\sim}{r}$ does no work in any deformation in which $\underset{\sim}{a}.\underset{\sim}{e}.\underset{\sim}{a} = 0$ and $\mathrm{tr}\,\underset{\sim}{e} = 0$. The extra-stress $\underset{\sim}{s}$ is indeterminate to within an arbitrary pressure and an arbitrary tension in the fibre direction. If, without loss of generality, we specify

$$\mathrm{tr}\,\underset{\sim}{s} = 0 \ , \qquad \underset{\sim}{a}.\underset{\sim}{s}.\underset{\sim}{a} = 0 \ , \tag{30}$$

then it follows from (28) that

$$\underset{\sim}{s} = 2\mu_T\underset{\sim}{e} + 2(\mu_L-\mu_T)(\underset{\sim}{a}\otimes\underset{\sim}{a}.\underset{\sim}{e}+\underset{\sim}{e}.\underset{\sim}{a}\otimes\underset{\sim}{a}) \ . \tag{31}$$

2.2 Linear elasticity - two families of fibres

Let us now consider a material which has linear elastic response and is reinforced by two families of fibres, with fibre directions $\underset{\sim}{a}$ and $\underset{\sim}{b}$. Suppose that the only anisotropic properties of the material are those which are due to the presence of the fibres, so that there are two preferred directions $\underset{\sim}{a}$ and $\underset{\sim}{b}$ at each point in the material.

If the two families of fibres are orthogonal (as in a cross-ply laminated material) then locally the material possesses material symmetry with respect to reflections in the planes normal to the fibres and so, locally, the material is *orthotropic* with respect to the planes normal to the fibres and the surfaces in which the fibres lie. If the two families are not necessarily orthogonal but are mechanically equivalent (i.e. are indistinguishable except for their directions, as in a balanced angle-ply laminate), then locally the material has material symmetry with respect to reflections in planes normal to the bisectors of the two families of fibres and again the material is locally orthotropic, but now with respect to the planes normal to the bisectors of the fibre families and the surfaces in which the fibres lie. However, in the first instance we shall not restrict ourselves to either of these special cases, and will consider the general case of reinforcement by two families of fibres which are not necessarily either orthogonal or mechanically equivalent.

By arguments similar to those used for a single fibre family, W is quadratic in $\mathbf{e}$, even in $\mathbf{a}$ and $\mathbf{b}$, and such that

$$W(\mathbf{Q}.\mathbf{e}.\mathbf{Q}^T,\ \mathbf{Q}.\mathbf{a},\ \mathbf{Q}.\mathbf{b}) = W(\mathbf{e},\mathbf{a},\mathbf{b})\ ,$$

where $\mathbf{Q}$ is any proper orthogonal tensor. It follows that W is an isotropic invariant of $\mathbf{e}$, $\mathbf{a}\otimes\mathbf{a}$ and $\mathbf{b}\otimes\mathbf{b}$. From tables of such invariants [1], the relations (6) and (7), and similar relations for $\mathbf{b}$, it follows that W is a function of

$$\begin{array}{llllll} \mathrm{tr}\,\mathbf{e}\,, & \mathrm{tr}\,\mathbf{e}^2\,, & \mathrm{tr}\,\mathbf{e}^3\,, & \mathbf{a}.\mathbf{e}.\mathbf{a}\,, & \mathbf{a}.\mathbf{e}^2.\mathbf{a}\,, & \mathbf{b}.\mathbf{e}.\mathbf{b}\,, \\ \mathbf{b}.\mathbf{e}^2.\mathbf{b}\,, & (\mathbf{a}.\mathbf{b})^2 = \cos^2 2\phi\,, & \cos 2\phi\ \mathbf{a}.\mathbf{e}.\mathbf{b}\,, & \cos 2\phi\ \mathbf{a}.\mathbf{e}^2.\mathbf{b}\,, & & \end{array} \tag{32}$$

where 2ϕ is the angle between the two fibre directions. However, with a substantial amount of algebra, we can prove the identity

$$\begin{aligned} \sin^2 2\phi\{(\mathrm{tr}\,\mathbf{e})^2 - \mathrm{tr}\,\mathbf{e}^2\} + 2\cos 2\phi\{(\mathbf{a}.\mathbf{e}.\mathbf{b})\,\mathrm{tr}\,\mathbf{e} - \mathbf{a}.\mathbf{e}^2.\mathbf{b}\} \\ - (\mathbf{a}.\mathbf{e}.\mathbf{a}+\mathbf{b}.\mathbf{e}.\mathbf{b})\,\mathrm{tr}\,\mathbf{e} + (\mathbf{a}.\mathbf{e}.\mathbf{a})(\mathbf{b}.\mathbf{e}.\mathbf{b}) \\ - (\mathbf{a}.\mathbf{e}.\mathbf{b})^2 + \mathbf{a}.\mathbf{e}^2.\mathbf{a} + \mathbf{b}.\mathbf{e}^2.\mathbf{b} = 0\,, \end{aligned} \tag{33}$$

and so one invariant, say $\cos 2\phi\ \mathbf{a}.\mathbf{e}^2.\mathbf{b}$, can be omitted from the list (32). It is then a straightforward matter to write down the most general function W which is a quadratic in $\mathbf{e}$ formed from the list (32) (less $\cos 2\phi\ \mathbf{a}.\mathbf{e}^2.\mathbf{b}$) and to determine $\boldsymbol{\sigma}$ from (4). This quadratic has thirteen elastic constants as coefficients. However, we shall now specialise to the cases which lead to orthotropic symmetry.

Orthogonal fibres. If the two families of fibres are orthogonal, then $\cos 2\phi = 0$, and the list (32) reduces to

$$\begin{array}{lllll} \mathrm{tr}\,\mathbf{e}\,, & \mathrm{tr}\,\mathbf{e}^2\,, & \mathrm{tr}\,\mathbf{e}^3\,, & \mathbf{a}.\mathbf{e}.\mathbf{a}\,, & \mathbf{a}.\mathbf{e}^2.\mathbf{a}\,, \\ \mathbf{b}.\mathbf{e}.\mathbf{b}\,, & \mathbf{b}.\mathbf{e}^2.\mathbf{b} & & & \end{array} \tag{34}$$

and the most general quadratic form for W is

$$W = \tfrac{1}{2}\lambda(\operatorname{tr}\underset{\sim}{e})^2 + \mu \operatorname{tr}\underset{\sim}{e}^2 + (\alpha_1 \underset{\sim}{a}.\underset{\sim}{e}.\underset{\sim}{a} + \alpha_2 \underset{\sim}{b}.\underset{\sim}{e}.\underset{\sim}{b}) \operatorname{tr}\underset{\sim}{e}$$

$$+ 2\mu_1 \underset{\sim}{a}.\underset{\sim}{e}^2.\underset{\sim}{a} + 2\mu_2 \underset{\sim}{b}.\underset{\sim}{e}^2.\underset{\sim}{b} + \tfrac{1}{2}\beta_1(\underset{\sim}{a}.\underset{\sim}{e}.\underset{\sim}{a})^2$$

$$+ \tfrac{1}{2}\beta_2(\underset{\sim}{b}.\underset{\sim}{e}.\underset{\sim}{b})^2 + \beta_3(\underset{\sim}{a}.\underset{\sim}{e}.\underset{\sim}{a})(\underset{\sim}{b}.\underset{\sim}{e}.\underset{\sim}{b}) , \qquad (35)$$

where λ, μ, α_1, α_2, μ_1, μ_2, β_1, β_2 and β_3 are elastic constants. The corresponding expression for the stress is

$$\underset{\sim}{\sigma} = (\lambda \operatorname{tr}\underset{\sim}{e} + \alpha_1 \underset{\sim}{a}.\underset{\sim}{e}.\underset{\sim}{a} + \alpha_2 \underset{\sim}{b}.\underset{\sim}{e}.\underset{\sim}{b})\,\underset{\sim}{I}$$

$$+ (\alpha_1 \operatorname{tr}\underset{\sim}{e} + \beta_1 \underset{\sim}{a}.\underset{\sim}{e}.\underset{\sim}{a} + \beta_3 \underset{\sim}{b}.\underset{\sim}{e}.\underset{\sim}{b})\,\underset{\sim}{a}\otimes\underset{\sim}{a}$$

$$+ (\alpha_2 \operatorname{tr}\underset{\sim}{e} + \beta_3 \underset{\sim}{a}.\underset{\sim}{e}.\underset{\sim}{a} + \beta_2 \underset{\sim}{b}.\underset{\sim}{e}.\underset{\sim}{b})\,\underset{\sim}{b}\otimes\underset{\sim}{b}$$

$$+ 2\mu\underset{\sim}{e} + 2\mu_1(\underset{\sim}{a}\otimes\underset{\sim}{a}.\underset{\sim}{e} + \underset{\sim}{e}.\underset{\sim}{a}\otimes\underset{\sim}{a}) + 2\mu_2(\underset{\sim}{b}\otimes\underset{\sim}{b}.\underset{\sim}{e} + \underset{\sim}{e}.\underset{\sim}{b}\otimes\underset{\sim}{b}) . \qquad (36)$$

If the coordinate axes are chosen so that $\underset{\sim}{a}$ has components (1,0,0) and $\underset{\sim}{b}$ has components (0,1,0), then (36) may be written as

$$\begin{pmatrix} \sigma_{11} \\ \sigma_{22} \\ \sigma_{33} \\ \sigma_{23} \\ \sigma_{31} \\ \sigma_{12} \end{pmatrix} = \begin{pmatrix} \begin{matrix}\lambda+2\alpha_1+\beta_1\\+2\mu+4\mu_1\end{matrix} & \begin{matrix}\lambda+\alpha_1\\+\alpha_2+\beta_3\end{matrix} & \lambda+\alpha_1 & 0 & 0 & 0 \\ \begin{matrix}\lambda+\alpha_1\\+\alpha_2+\beta_3\end{matrix} & \begin{matrix}\lambda+2\alpha_2+\beta_2\\+2\mu+4\mu_2\end{matrix} & \lambda+\alpha_2 & 0 & 0 & 0 \\ \lambda+\alpha_1 & \lambda+\alpha_2 & \lambda+2\mu & 0 & 0 & 0 \\ 0 & 0 & 0 & 2(\mu+\mu_2) & 0 & 0 \\ 0 & 0 & 0 & 0 & 2(\mu+\mu_1) & 0 \\ 0 & 0 & 0 & 0 & 0 & 2(\mu+\mu_1+\mu_2) \end{pmatrix} \begin{pmatrix} e_{11} \\ e_{22} \\ e_{33} \\ e_{23} \\ e_{31} \\ e_{12} \end{pmatrix} \qquad (37)$$

This is of the usual form for an orthotropic material, and enables the elastic constants λ, μ, α_1, α_2, μ_1, μ_2, β_1, β_2 and β_3 to be related to elastic constants with more direct physical interpretations; for example, we see that $\mu+\mu_2$, $\mu+\mu_1$, $\mu+\mu_1+\mu_2$ are shear moduli for shear on the planes normal to and parallel to the fibres. There are nine independent elastic constants.

If the material is incompressible then $\mathrm{tr}\,\underset{\sim}{e} = 0$ and the terms involving λ, α_1 and α_2 are omitted from (35) and (36), and a reaction stress in the form of a hydrostatic pressure $-p\underset{\sim}{I}$ is added on the right of (36). If the material is inextensible in both fibre directions, then $\underset{\sim}{a}.\underset{\sim}{e}.\underset{\sim}{a} = 0$ and $\underset{\sim}{b}.\underset{\sim}{e}.\underset{\sim}{b} = 0$. Then W takes the form

$$W = \tfrac{1}{2}\lambda(\mathrm{tr}\,\underset{\sim}{e})^2 + \mu\,\mathrm{tr}\,\underset{\sim}{e}^2 + 2\mu_1\underset{\sim}{a}.\underset{\sim}{e}^2.\underset{\sim}{a} + 2\mu_2\underset{\sim}{b}.\underset{\sim}{e}^2.\underset{\sim}{b} + T_a\underset{\sim}{a}.\underset{\sim}{e}.\underset{\sim}{a} + T_b\underset{\sim}{b}.\underset{\sim}{e}.\underset{\sim}{b}\,, \qquad (38)$$

where the Lagrangian multipliers T_a and T_b represent arbitrary tensions in the two fibre directions. Then $\underset{\sim}{\sigma} = \underset{\sim}{s}+\underset{\sim}{r}$, where the reaction stress $\underset{\sim}{r}$ is given by

$$\underset{\sim}{r} = T_a\underset{\sim}{a}\otimes\underset{\sim}{a} + T_b\underset{\sim}{b}\otimes\underset{\sim}{b}\,. \qquad (39)$$

The extra-stress $\underset{\sim}{s}$ is given by

$$\underset{\sim}{s} = \lambda\,\mathrm{tr}\,\underset{\sim}{e}(\underset{\sim}{I}-\underset{\sim}{a}\otimes\underset{\sim}{a}-\underset{\sim}{b}\otimes\underset{\sim}{b}) + 2\mu\underset{\sim}{e} + 2\mu_1(\underset{\sim}{a}\otimes\underset{\sim}{a}.\underset{\sim}{e}+\underset{\sim}{e}.\underset{\sim}{a}\otimes\underset{\sim}{a}) + 2\mu_2(\underset{\sim}{b}\otimes\underset{\sim}{b}.\underset{\sim}{e}+\underset{\sim}{e}.\underset{\sim}{b}\otimes\underset{\sim}{b})\,, \qquad (40)$$

and has been chosen so that $\underset{\sim}{a}.\underset{\sim}{s}.\underset{\sim}{a} = 0$ and $\underset{\sim}{b}.\underset{\sim}{s}.\underset{\sim}{b} = 0$. The number of independent elastic constants is reduced to four.

If the material is both incompressible and inextensible, then

$$\underset{\sim}{r} = T_a\underset{\sim}{a}\otimes\underset{\sim}{a} + T_b\underset{\sim}{b}\otimes\underset{\sim}{b} - p\underset{\sim}{I}\,, \qquad (41)$$

$$\underset{\sim}{s} = 2\mu\underset{\sim}{e} + 2\mu_1(\underset{\sim}{a}\otimes\underset{\sim}{a}.\underset{\sim}{e}+\underset{\sim}{e}.\underset{\sim}{a}\otimes\underset{\sim}{a}) + 2\mu_2(\underset{\sim}{b}\otimes\underset{\sim}{b}.\underset{\sim}{e}+\underset{\sim}{e}.\underset{\sim}{b}\otimes\underset{\sim}{b})\,, \qquad (42)$$

where $\underset{\sim}{s}$ has been chosen so that $\mathrm{tr}\,\underset{\sim}{s} = 0$, $\underset{\sim}{a}.\underset{\sim}{s}.\underset{\sim}{a} = 0$ and $\underset{\sim}{b}.\underset{\sim}{s}.\underset{\sim}{b} = 0$. There are now three independent elastic constants.

Two mechanically equivalent families of fibres. If the two families of fibres are mechanically equivalent, then W must be symmetric with respect to interchanges of $\underset{\sim}{a}$ and $\underset{\sim}{b}$. Hence the set (32) (with $\cos 2\phi\,\underset{\sim}{a}.\underset{\sim}{e}^2.\underset{\sim}{b}$ omitted) is replaced by

$$\mathrm{tr}\,\underset{\sim}{e}\,,\quad \mathrm{tr}\,\underset{\sim}{e}^2\,,\quad \mathrm{tr}\,\underset{\sim}{e}^3\,,\quad \underset{\sim}{a}.\underset{\sim}{e}.\underset{\sim}{a}+\underset{\sim}{b}.\underset{\sim}{e}.\underset{\sim}{b}\,,\quad (\underset{\sim}{a}.\underset{\sim}{e}.\underset{\sim}{a})(\underset{\sim}{b}.\underset{\sim}{e}.\underset{\sim}{b})\,,$$
$$\underset{\sim}{a}.\underset{\sim}{e}^2.\underset{\sim}{a}+\underset{\sim}{b}.\underset{\sim}{e}^2.\underset{\sim}{b}\,,\quad (\underset{\sim}{a}.\underset{\sim}{e}^2.\underset{\sim}{a})(\underset{\sim}{b}.\underset{\sim}{e}^2.\underset{\sim}{b})\,,\quad \cos 2\phi\,\underset{\sim}{a}.\underset{\sim}{e}.\underset{\sim}{b}\,,\quad \cos^2 2\phi\,. \qquad (43)$$

The most general expression for W which is quadratic in $\underset{\sim}{e}$ is now

$$W = \tfrac{1}{2}\lambda(\mathrm{tr}\,\underset{\sim}{e})^2 + \mu\,\mathrm{tr}\,\underset{\sim}{e}^2 + \gamma_1\{(\underset{\sim}{a}.\underset{\sim}{e}.\underset{\sim}{a})^2 + (\underset{\sim}{b}.\underset{\sim}{e}.\underset{\sim}{b})^2\}$$
$$+ \gamma_2(\underset{\sim}{a}.\underset{\sim}{e}.\underset{\sim}{b})^2 + \gamma_3(\underset{\sim}{a}.\underset{\sim}{e}.\underset{\sim}{a}+\underset{\sim}{b}.\underset{\sim}{e}.\underset{\sim}{b})\,\mathrm{tr}\,\underset{\sim}{e}$$
$$+ \gamma_4\cos 2\phi(\underset{\sim}{a}.\underset{\sim}{e}.\underset{\sim}{b})\,\mathrm{tr}\,\underset{\sim}{e} + \gamma_5\cos 2\phi(\underset{\sim}{a}.\underset{\sim}{e}.\underset{\sim}{a}+\underset{\sim}{b}.\underset{\sim}{e}.\underset{\sim}{b})(\underset{\sim}{a}.\underset{\sim}{e}.\underset{\sim}{b})$$
$$+ \gamma_6(\underset{\sim}{a}.\underset{\sim}{e}.\underset{\sim}{a})(\underset{\sim}{b}.\underset{\sim}{e}.\underset{\sim}{b}) + \gamma_7(\underset{\sim}{a}.\underset{\sim}{e}^2.\underset{\sim}{a}+\underset{\sim}{b}.\underset{\sim}{e}^2.\underset{\sim}{b}), \tag{44}$$

where the nine coefficients λ, μ, $\gamma_1,\ldots,\gamma_7$ are even functions of $\cos 2\phi$. This leads to the constitutive equation

$$\underset{\sim}{\sigma} = \{\lambda\,\mathrm{tr}\,\underset{\sim}{e} + \gamma_3(\underset{\sim}{a}.\underset{\sim}{e}.\underset{\sim}{a}+\underset{\sim}{b}.\underset{\sim}{e}.\underset{\sim}{b}) + \gamma_4\underset{\sim}{a}.\underset{\sim}{e}.\underset{\sim}{b}\cos 2\phi\}\underset{\sim}{I}$$
$$+ 2\mu\underset{\sim}{e} + \{\gamma_3\,\mathrm{tr}\,\underset{\sim}{e} + 2\gamma_1\underset{\sim}{a}.\underset{\sim}{e}.\underset{\sim}{a}+\gamma_6\underset{\sim}{b}.\underset{\sim}{e}.\underset{\sim}{b}+\gamma_5\underset{\sim}{a}.\underset{\sim}{e}.\underset{\sim}{b}\cos 2\phi\}\underset{\sim}{a}\otimes\underset{\sim}{a}$$
$$+ \{\gamma_3\,\mathrm{tr}\,\underset{\sim}{e} + \gamma_6\underset{\sim}{a}.\underset{\sim}{e}.\underset{\sim}{a}+2\gamma_1\underset{\sim}{b}.\underset{\sim}{e}.\underset{\sim}{b}+\gamma_5\underset{\sim}{a}.\underset{\sim}{e}.\underset{\sim}{b}\cos 2\phi\}\underset{\sim}{b}\otimes\underset{\sim}{b}$$
$$+ [\tfrac{1}{2}\cos 2\phi\{\gamma_4\,\mathrm{tr}\,\underset{\sim}{e} + \gamma_5(\underset{\sim}{a}.\underset{\sim}{e}.\underset{\sim}{a}+\underset{\sim}{b}.\underset{\sim}{e}.\underset{\sim}{b})\} + \gamma_2\underset{\sim}{a}.\underset{\sim}{e}.\underset{\sim}{b}](\underset{\sim}{a}\otimes\underset{\sim}{b}+\underset{\sim}{b}\otimes\underset{\sim}{a})$$
$$+ \gamma_7(\underset{\sim}{a}\otimes\underset{\sim}{a}.\underset{\sim}{e}+\underset{\sim}{e}.\underset{\sim}{a}\otimes\underset{\sim}{a}+\underset{\sim}{b}\otimes\underset{\sim}{b}.\underset{\sim}{e}+\underset{\sim}{e}.\underset{\sim}{b}\otimes\underset{\sim}{b}). \tag{45}$$

In many cases it is advantageous to express the equations in terms of the bisectors of the fibre directions, because these bisectors are mutually orthogonal. For this we introduce unit vectors $\underset{\sim}{c}$ and $\underset{\sim}{d}$, where

$$\underset{\sim}{c} = \tfrac{1}{2}(\underset{\sim}{a}+\underset{\sim}{b})/\cos\phi, \qquad \underset{\sim}{d} = \tfrac{1}{2}(\underset{\sim}{a}-\underset{\sim}{b})/\sin\phi,$$
$$\underset{\sim}{a} = \underset{\sim}{c}\cos\phi + \underset{\sim}{d}\sin\phi, \qquad \underset{\sim}{b} = \underset{\sim}{c}\cos\phi - \underset{\sim}{d}\sin\phi. \tag{46}$$

On substituting for $\underset{\sim}{a}$ and $\underset{\sim}{b}$ from (46) into (44), we obtain an expression for W of the same form as (35), with $\underset{\sim}{a}$ and $\underset{\sim}{b}$ replaced by $\underset{\sim}{c}$ and $\underset{\sim}{d}$ respectively, and with the coefficients becoming functions of $\cos 2\phi$. Hence this case also corresponds to orthotropic symmetry, and if $\underset{\sim}{c}$ and $\underset{\sim}{d}$ are chosen to lie in the x_1 and x_2 coordinate directions, the constitutive equation assumes the form (37). There are again nine independent elastic coefficients, which are functions of $\cos 2\phi$. The constitutive equation in terms of $\underset{\sim}{c}$ and $\underset{\sim}{d}$ is obtained by substituting for $\underset{\sim}{a}$ and $\underset{\sim}{b}$ from (46) into (45). The expression is obviously complicated. An alternative and rather simpler procedure is to observe from the beginning that W can be expressed as a function of $\underset{\sim}{e}$, $\underset{\sim}{c}$, $\underset{\sim}{d}$ and $\cos 2\phi$, with $\underset{\sim}{c}.\underset{\sim}{d} = 0$. Hence it is possible to proceed as in the case of orthogonal fibres, with $\underset{\sim}{a}$ and $\underset{\sim}{b}$ replaced by $\underset{\sim}{c}$ and $\underset{\sim}{d}$, and the coefficients in the expression for W regarded as functions of $\cos 2\phi$. However, with the alternative procedure it is less easy to proceed to the case of fibre inextensibility.

If the material is incompressible then $\mathrm{tr}\,\underset{\sim}{e} = 0$ and the terms involving λ, γ_3 and γ_4 are omitted from (44) and (45), and a reaction stress $-p\underset{\sim}{I}$ is added. If the material is inextensible in both fibre directions, then $\underset{\sim}{a}.\underset{\sim}{e}.\underset{\sim}{a} = 0$ and $\underset{\sim}{b}.\underset{\sim}{e}.\underset{\sim}{b} = 0$, and (45) is replaced by

$$\underset{\sim}{\sigma} = \lambda \underset{\sim}{I}\,\mathrm{tr}\,\underset{\sim}{e} + 2\mu\underset{\sim}{e} + (\tfrac{1}{2}\gamma_4 \cos 2\phi\,\mathrm{tr}\,\underset{\sim}{e} + \gamma_2 \underset{\sim}{a}.\underset{\sim}{e}.\underset{\sim}{b})(\underset{\sim}{a}\otimes\underset{\sim}{b}+\underset{\sim}{b}\otimes\underset{\sim}{a})$$

$$+ \gamma_7(\underset{\sim}{a}\otimes\underset{\sim}{a}.\underset{\sim}{e}+\underset{\sim}{e}.\underset{\sim}{a}\otimes\underset{\sim}{a}+\underset{\sim}{b}\otimes\underset{\sim}{b}.\underset{\sim}{e}+\underset{\sim}{e}.\underset{\sim}{b}\otimes\underset{\sim}{b}) + T_a\underset{\sim}{a}\otimes\underset{\sim}{a} + T_b\underset{\sim}{b}\otimes\underset{\sim}{b}\,. \tag{47}$$

The number of independent elastic constants is reduced to five, and the last two terms represent the reaction stress. In terms of the vectors $\underset{\sim}{c}$ and $\underset{\sim}{d}$, (47) is

$$\underset{\sim}{\sigma} = \lambda\underset{\sim}{I}\,\mathrm{tr}\,\underset{\sim}{e} + 2\mu\underset{\sim}{e} + \{\tfrac{1}{2}\gamma_4 \cos 2\phi\,\mathrm{tr}\,\underset{\sim}{e} + 2\gamma_2(\underset{\sim}{c}.\underset{\sim}{e}.\underset{\sim}{c}\cos^2\phi + \underset{\sim}{d}.\underset{\sim}{e}.\underset{\sim}{d}\sin^2\phi\} \times$$

$$\times(\underset{\sim}{c}\otimes\underset{\sim}{c}\cos^2\phi - \underset{\sim}{d}\otimes\underset{\sim}{d}\sin^2\phi) + \gamma_7\{\cos^2\phi\,(\underset{\sim}{c}\otimes\underset{\sim}{c}.\underset{\sim}{e}+\underset{\sim}{e}.\underset{\sim}{c}\otimes\underset{\sim}{c})$$

$$+ \sin^2\phi\,(\underset{\sim}{d}\otimes\underset{\sim}{d}.\underset{\sim}{e}+\underset{\sim}{e}.\underset{\sim}{d}\otimes\underset{\sim}{d})\} + (T_a+T_b)(\underset{\sim}{c}\otimes\underset{\sim}{c}\cos^2\phi + \underset{\sim}{d}\otimes\underset{\sim}{d}\sin^2\phi)$$

$$+ (T_a-T_b)\sin\phi\cos\phi\,(\underset{\sim}{c}\otimes\underset{\sim}{d}+\underset{\sim}{d}\otimes\underset{\sim}{a})\,. \tag{48}$$

If the material is incompressible as well as inextensible in the two fibre directions, then $\mathrm{tr}\,\underset{\sim}{e} = 0$, and there is an additional reaction stress $-p\underset{\sim}{I}$. There are only three independent elastic constants, and (47) and (48) reduce to

$$\underset{\sim}{\sigma} = 2\mu\underset{\sim}{e} + \gamma_2\underset{\sim}{a}.\underset{\sim}{e}.\underset{\sim}{b}(\underset{\sim}{a}\otimes\underset{\sim}{b}+\underset{\sim}{b}\otimes\underset{\sim}{a})$$

$$+ \gamma_7(\underset{\sim}{a}\otimes\underset{\sim}{a}.\underset{\sim}{e}+\underset{\sim}{e}.\underset{\sim}{a}\otimes\underset{\sim}{a}+\underset{\sim}{b}\otimes\underset{\sim}{b}.\underset{\sim}{e}+\underset{\sim}{e}.\underset{\sim}{b}\otimes\underset{\sim}{b}) + T_a\underset{\sim}{a}\otimes\underset{\sim}{a} + T_b\underset{\sim}{b}\otimes\underset{\sim}{b} - p\underset{\sim}{I}\,, \tag{49}$$

$$\underset{\sim}{\sigma} = 2\mu\underset{\sim}{e} + 2\gamma_2(\underset{\sim}{c}.\underset{\sim}{e}.\underset{\sim}{c}\cos^2\phi + \underset{\sim}{d}.\underset{\sim}{e}.\underset{\sim}{d}\sin^2\phi)(\underset{\sim}{c}\otimes\underset{\sim}{c}\cos^2\phi - \underset{\sim}{d}\otimes\underset{\sim}{d}\sin^2\phi)$$
$$+ \gamma_7\{\cos^2\phi(\underset{\sim}{c}\otimes\underset{\sim}{c}.\underset{\sim}{e}+\underset{\sim}{e}.\underset{\sim}{c}\otimes\underset{\sim}{c}) + \sin^2\phi(\underset{\sim}{d}\otimes\underset{\sim}{d}.\underset{\sim}{e}+\underset{\sim}{e}.\underset{\sim}{d}\otimes\underset{\sim}{d})\}$$
$$+ (T_a+T_b)(\underset{\sim}{c}\otimes\underset{\sim}{c}\cos^2\phi + \underset{\sim}{d}\otimes\underset{\sim}{d}\sin^2\phi)$$
$$+ (T_a-T_b)\sin\phi\cos\phi(\underset{\sim}{c}\otimes\underset{\sim}{d}+\underset{\sim}{d}\otimes\underset{\sim}{c}) - p\underset{\sim}{I}. \tag{50}$$

In all of the cases discussed above, it is possible to express $\underset{\sim}{\sigma}$ as $\underset{\sim}{\sigma} = \underset{\sim}{r}+\underset{\sim}{s}$, where $\underset{\sim}{r}$ is the reaction stress, and without loss of generality $\underset{\sim}{s}$ may be chosen so that $\underset{\sim}{a}.\underset{\sim}{s}.\underset{\sim}{a} = 0$, $\underset{\sim}{b}.\underset{\sim}{s}.\underset{\sim}{b} = 0$ when the material is inextensible in the directions $\underset{\sim}{a}$ and $\underset{\sim}{b}$, and $\mathrm{tr}\,\underset{\sim}{s} = 0$ when the material is incompressible.

3. FINITE ELASTIC CONSTITUTIVE EQUATIONS FOR FIBRE-REINFORCED MATERIAL

3.1 Kinematics of finite deformations

So far we have considered only small deformations, for which it is not necessary to distinguish between the fibre directions in the undeformed and deformed configurations of a body. We now turn to finite deformations.

The deformation will be referred to a fixed frame of reference, and to rectangular cartesian coordinates in this frame. We consider a body which is initially in a *reference configuration* in which a typical particle has position vector $\underset{\sim}{X}$, with components X_R. At a subsequent time the body is in a *deformed configuration*, and the generic particle has position vector $\underset{\sim}{x}$, with coordinates x_i. Thus the deformation is described by equations of the form

$$\underset{\sim}{x} = \underset{\sim}{x}(\underset{\sim}{X}), \qquad \text{or} \qquad x_i = x_i(X_R) \tag{51}$$

which give the spatial coordinates x_i in terms of the material coordinates X_R. The deformation gradient tensor $\underset{\sim}{F}$ has cartesian coordinates F_{iR}, where

$$F_{iR} = \partial x_i/\partial X_R . \tag{52}$$

If δV and δv are volumes of a material volume element in the reference and deformed configurations respectively, and ρ_0 and ρ the densities of the element in these two configurations, then

$$\frac{\delta v}{\delta V} = \frac{\rho_0}{\rho} = \det \underset{\sim}{F} . \tag{53}$$

We shall also employ the deformation tensors $\underset{\sim}{C}$ and $\underset{\sim}{B}$, with cartesian components C_{RS} and B_{ij} respectively, where

$$\underset{\sim}{C} = \underset{\sim}{F}^T.\underset{\sim}{F} , \qquad \underset{\sim}{B} = \underset{\sim}{F}.\underset{\sim}{F}^T , \tag{54}$$

$$C_{RS} = \frac{\partial x_i}{\partial X_R}\frac{\partial x_i}{\partial X_S} = F_{iR}F_{iS} , \qquad B_{ij} = \frac{\partial x_i}{\partial X_R}\frac{\partial x_j}{\partial X_R} = F_{iR}F_{jR} . \tag{55}$$

Suppose that in the reference configuration a fibre direction is defined by a unit vector field $\underset{\sim}{a}_0(\underset{\sim}{X})$ with cartesian components $a_R^{(0)}$. In a deformation the fibres, being material line elements, will be convected with the particles of the body, so that in the deformed configuration the fibre direction may be described by a unit vector field $\underset{\sim}{a}(\underset{\sim}{x})$ with cartesian components a_i. In general the fibres will also stretch; suppose that a fibre element has length δL in the reference configuration and length $\delta\ell$ in the deformed configuration. Then the *stretch* $\lambda = \delta\ell/\delta L$.

Consider a fibre element whose ends have coordinates X_R and $X_R + a_R^{(0)}\delta L$ in the reference configuration and x_i and $x_i + a_i\delta\ell$ in the deformed configuration. Then, from (51),

$$x_i = x_i(X_R) , \qquad x_i + a_i\delta\ell = x_i(X_R + a_R^{(0)}\delta L)$$

and it follows that

$$a_i\delta\ell = \frac{\partial x_i}{\partial X_R} a_R^{(0)}\delta L .$$

Thus

$$\lambda a_i = \frac{\partial x_i}{\partial X_R} a_R^{(0)} , \qquad \text{or} \qquad \lambda\underset{\sim}{a} = \underset{\sim}{F}.\underset{\sim}{a}_0 . \tag{56}$$

This relates the fibre directions in the reference and deformed configurations. Also, since $\underset{\sim}{a}$ is a unit vector,

$$\lambda^2 a_i a_i = \lambda^2 = \frac{\partial x_i}{\partial X_R}\frac{\partial x_i}{\partial X_S} a_R^{(0)} a_S^{(0)} = C_{RS} a_R^{(0)} a_S^{(0)} = \underset{\sim}{a}_0 . \underset{\sim}{C} . \underset{\sim}{a}_0 , \tag{57}$$

which determines the fibre stretch.

3.2 Finite elasticity for one family of fibres

We consider a finite elastic solid with a strain energy W which is a function of the deformation gradients F_{iR}. Then by standard arguments in the development of finite elasticity theory (for example [4]), which are in no way affected by the presence of one or more families of fibres, W can be expressed as a function of the components C_{RS}, and the constitutive equation for the stress is

$$\sigma_{ij} = \frac{\rho}{\rho_0} F_{iR} F_{jS} \left\{ \frac{\partial W}{\partial C_{RS}} + \frac{\partial W}{\partial C_{SR}} \right\} . \tag{58}$$

Consider a material reinforced by a single family of fibres with initial fibre direction $\underset{\sim}{a}_0$. Then by arguments similar to those used in the linear elastic case, W can be expressed as a function of $\underset{\sim}{C}$ and $\underset{\sim}{a}_0 \otimes \underset{\sim}{a}_0$.

Now select a new reference configuration which is obtained by a rigid rotation of the undeformed material and the fibres, so that a typical particle is at $\bar{\underset{\sim}{X}} = \underset{\sim}{Q} . \underset{\sim}{X}$ and the fibre direction is $\underset{\sim}{Q} . \underset{\sim}{a}_0$, where $\underset{\sim}{Q}$ is a proper orthogonal tensor. The deformation tensor from the new reference configuration is $\bar{\underset{\sim}{C}} = \underset{\sim}{Q} . \underset{\sim}{C} . \underset{\sim}{Q}^T$. However, this change of reference configuration leaves W unaltered, and so

$$W(\underset{\sim}{C}, \underset{\sim}{a}_0 \otimes \underset{\sim}{a}_0) = W(\underset{\sim}{Q} . \underset{\sim}{C} . \underset{\sim}{Q}^T, \underset{\sim}{Q} . \underset{\sim}{a}_0 \otimes \underset{\sim}{a}_0 . \underset{\sim}{Q}^T) \tag{59}$$

for all proper orthogonal tensors $\underset{\sim}{Q}$. Hence W is an isotropic invariant of $\underset{\sim}{C}$ and $\underset{\sim}{a}_0 \otimes \underset{\sim}{a}_0$. Therefore, by arguments similar to those used in deriving the set (8), with $\underset{\sim}{e}$ and $\underset{\sim}{a}$ replaced by $\underset{\sim}{C}$ and $\underset{\sim}{a}_0$ respectively, it follows that W can be expressed as a function of the invariants

$$I_1 = \mathrm{tr}\,\underset{\sim}{C}, \qquad I_2 = \tfrac{1}{2}\{(\mathrm{tr}\,\underset{\sim}{C})^2 - \mathrm{tr}\,\underset{\sim}{C}^2\}, \qquad I_3 = \det \underset{\sim}{C} = (\rho_0/\rho)^2,$$

$$I_4 = \underset{\sim}{a}_0 \cdot \underset{\sim}{C} \cdot \underset{\sim}{a}_0 = \lambda^2, \qquad I_5 = \underset{\sim}{a}_0 \cdot \underset{\sim}{C}^2 \cdot \underset{\sim}{a}_0, \tag{60}$$

where, for convenience, $\mathrm{tr}\,\underset{\sim}{C}$, $\mathrm{tr}\,\underset{\sim}{C}^2$ and $\mathrm{tr}\,\underset{\sim}{C}^3$ have been replaced by the equivalent set I_1, I_2 and I_3.

From (58) it then follows that

$$\sigma_{ij} = I_3^{-\frac{1}{2}} F_{iR} F_{jS} \sum_{\alpha=1}^{5} W_\alpha \left(\frac{\partial I_\alpha}{C_{RS}} + \frac{\partial I_\alpha}{C_{SR}} \right), \tag{61}$$

where W_α denotes $\partial W/\partial I_\alpha$. From (60) we obtain

$$\frac{\partial I_1}{\partial C_{RS}} = \delta_{RS}, \qquad \frac{\partial I_2}{\partial C_{RS}} = I_1 \delta_{RS} - C_{RS},$$

$$\frac{\partial I_3}{\partial C_{RS}} = I_2 \delta_{RS} - I_1 C_{RS} + C_{RP} C_{PS},$$

$$\frac{\partial I_4}{\partial C_{RS}} = a_R^{(0)} a_S^{(0)}, \qquad \frac{\partial I_5}{\partial C_{RS}} = 2 a_R^{(0)} a_P^{(0)} C_{PS}. \tag{62}$$

Hence (61) can be written as

$$\underset{\sim}{\sigma} = 2 I_3^{-\frac{1}{2}} \underset{\sim}{F} . \{ (W_1 + I_1 W_2 + I_2 W_3) \underset{\sim}{I} - (W_2 + I_1 W_3) \underset{\sim}{C}$$

$$+ W_3 \underset{\sim}{C}^2 + W_4 \underset{\sim}{a}_0 \otimes \underset{\sim}{a}_0 + W_5 (\underset{\sim}{a}_0 \otimes \underset{\sim}{C} . \underset{\sim}{a}_0 + \underset{\sim}{a}_0 . \underset{\sim}{C} \otimes \underset{\sim}{a}_0) \} . \underset{\sim}{F}^T .$$

Using (54), (56) and (60), this becomes

$$\underset{\sim}{\sigma} = 2 I_3^{-\frac{1}{2}} \{ (W_1 + I_1 W_2 + I_2 W_3) \underset{\sim}{B} - (W_2 + I_1 W_3) \underset{\sim}{B}^2$$

$$+ W_3 \underset{\sim}{B}^3 + I_4 W_4 \underset{\sim}{a} \otimes \underset{\sim}{a} + I_4 W_5 (\underset{\sim}{a} \otimes \underset{\sim}{B} . \underset{\sim}{a} + \underset{\sim}{a} . \underset{\sim}{B} \otimes \underset{\sim}{a}) \} . \tag{63}$$

Further simplification follows by using the Cayley-Hamilton theorem for $\underset{\sim}{B}$, namely

$$\underset{\sim}{B}^3 - I_1 \underset{\sim}{B}^2 + I_2 \underset{\sim}{B} - I_3 \underset{\sim}{I} = \underset{\sim}{O}, \tag{64}$$

and, since $\det \underset{\sim}{B} \neq 0$, the relation

$$\underset{\sim}{B}^2 - I_1 \underset{\sim}{B} + I_2 \underset{\sim}{I} - I_3 \underset{\sim}{B}^{-1} = \underset{\sim}{O}, \tag{65}$$

to eliminate $\underset{\sim}{B}^3$ and $\underset{\sim}{B}^2$ from (63) in favour of $\underset{\sim}{B}^{-1}$, which gives

$$\underset{\sim}{\sigma} = 2I_3^{-\frac{1}{2}}\{(I_2W_2+I_3W_3)\underset{\sim}{I} + W_1\underset{\sim}{B} - I_3W_2\underset{\sim}{B}^{-1} + I_4W_4\underset{\sim}{a}\otimes\underset{\sim}{a} + I_4W_5(\underset{\sim}{a}\otimes\underset{\sim}{B}.\underset{\sim}{a}+\underset{\sim}{a}.\underset{\sim}{B}\otimes\underset{\sim}{a})\} . \tag{66}$$

This is equivalent to results given by Ericksen and Rivlin [5] for transversely isotropic elastic materials.

If the material is incompressible, then $I_3 = 1$, W is a function of I_1, I_2, I_4 and I_5, but a term $-\frac{1}{2}p(I_3-1)$, where p is a Lagrangian multiplier, may be added to W. This leads to the constitutive equation

$$\underset{\sim}{\sigma} = 2\{W_1\underset{\sim}{B} - W_2\underset{\sim}{B}^{-1} + I_4W_4\underset{\sim}{a}\otimes\underset{\sim}{a} + I_4W_5(\underset{\sim}{a}\otimes\underset{\sim}{B}.\underset{\sim}{a}+\underset{\sim}{a}.\underset{\sim}{B}\otimes\underset{\sim}{a})\} - p\underset{\sim}{I} , \tag{67}$$

and p is a reaction pressure.

If, in addition, the material is inextensible in the fibre direction $\underset{\sim}{a}$, then $I_4 = \lambda^2 = 1$, W depends on I_1, I_2 and I_5, and a term $\frac{1}{2}T(I_4-1)$ may be added to W, where T is another Lagrangian multiplier. Then

$$\underset{\sim}{\sigma} = 2\{W_1\underset{\sim}{B} - W_2\underset{\sim}{B}^{-1} + W_5(\underset{\sim}{a}\otimes\underset{\sim}{B}.\underset{\sim}{a}+\underset{\sim}{a}.\underset{\sim}{B}\otimes\underset{\sim}{a})\} - p\underset{\sim}{I} + T\underset{\sim}{a}\otimes\underset{\sim}{a} , \tag{68}$$

and so again T is identified as an arbitrary fibre tension which is a reaction to the inextensibility constraint.

3.3 Finite elasticity for two families of fibres

If an elastic body is reinforced by two families of fibres, whose directions in the reference configuration are defined by unit vector fields $\underset{\sim}{a}_0$ and $\underset{\sim}{b}_0$ respectively, and in the deformed configuration by $\underset{\sim}{a}$ and $\underset{\sim}{b}$ respectively, then, by arguments similar to those used above, the strain-energy function W is an isotropic invariant of $\underset{\sim}{C}$, $\underset{\sim}{a}_0\otimes\underset{\sim}{a}_0$ and $\underset{\sim}{b}_0\otimes\underset{\sim}{b}_0$. It follows by analogy with (32) and (33) that W can be expressed as a function of

$$I_1 , \quad I_2 , \quad I_3 , \quad I_4 , \quad I_5 , \quad I_6 = \underset{\sim}{b}_0.\underset{\sim}{C}.\underset{\sim}{b}_0 ,$$

$$I_7 = \underset{\sim}{b}_0.\underset{\sim}{C}^2.\underset{\sim}{b}_0 , \quad I_8 = \cos 2\Phi\, \underset{\sim}{a}_0.\underset{\sim}{C}.\underset{\sim}{b}_0 \quad \text{and} \quad \cos^2 2\Phi , \tag{69}$$

where $\cos 2\Phi = \underset{\sim}{a}_0.\underset{\sim}{b}_0$ is the cosine of the angle between the two families of fibres in the reference configuration. The angle 2ϕ between the families of fibres in the deformed configuration is given by

$$\cos 2\phi = \underset{\sim}{a}.\underset{\sim}{b} = (I_4 I_6)^{-\frac{1}{2}} \underset{\sim}{a}_0.\underset{\sim}{C}.\underset{\sim}{b}_0 . \tag{70}$$

If the families of fibres are orthogonal in the reference configuration, then the material is orthotropic in this configuration; and W is a function of $I_1,\ldots,I_7$. Then, by arguments similar to those which lead to (66), the constitutive equation may be written as

$$\underset{\sim}{\sigma} = 2I_3^{-\frac{1}{2}}\{(I_2W_2+I_3W_3)\underset{\sim}{I} + W_1\underset{\sim}{B} - I_3W_2\underset{\sim}{B}^{-1} + I_4W_4\underset{\sim}{a}\otimes\underset{\sim}{a}$$

$$+ I_6W_6\underset{\sim}{b}\otimes\underset{\sim}{b} + I_4W_5(\underset{\sim}{a}\otimes\underset{\sim}{B}.\underset{\sim}{a}+\underset{\sim}{a}.\underset{\sim}{B}\otimes\underset{\sim}{a}) + I_6W_7(\underset{\sim}{b}\otimes\underset{\sim}{B}.\underset{\sim}{b}+\underset{\sim}{b}.\underset{\sim}{B}\otimes\underset{\sim}{b})\} . \tag{71}$$

This is in agreement with results for orthotropic elastic materials given by Smith and Rivlin [6] and Green and Adkins [7]. If, in addition, the material is incompressible, then $I_3 = 1$ and the term $(I_2W_2+I_3W_3)\underset{\sim}{I}$ in (71) is replaced by $-\frac{1}{2}p\underset{\sim}{I}$, where p is a reaction pressure. If, furthermore, the material is inextensible in the two fibre directions, then $I_4 = 1$, $I_6 = 1$, and (71) is replaced by

$$\underset{\sim}{\sigma} = 2\{W_1\underset{\sim}{B} - W_2\underset{\sim}{B}^{-1} + W_5(\underset{\sim}{a}\otimes\underset{\sim}{B}.\underset{\sim}{a}+\underset{\sim}{a}.\underset{\sim}{B}\otimes\underset{\sim}{a}) + W_7(\underset{\sim}{b}\otimes\underset{\sim}{B}.\underset{\sim}{b}+\underset{\sim}{b}.\underset{\sim}{B}\otimes\underset{\sim}{b})\}$$

$$- p\underset{\sim}{I} + T_a\underset{\sim}{a}\otimes\underset{\sim}{a} + T_b\underset{\sim}{b}\otimes\underset{\sim}{b} , \tag{72}$$

where the last three terms represent the reaction stress, and T_a and T_b are arbitrary fibre tensions.

If the two families of fibres are mechanically equivalent, then the material is locally orthotropic in the reference configuration with respect to the planes which bisect $\underset{\sim}{a}_0$ and $\underset{\sim}{b}_0$ and the planes containing $\underset{\sim}{a}_0$ and $\underset{\sim}{b}_0$. Then W is a function of $I_1,\ldots,I_8$ and symmetric with respect to interchanges of $\underset{\sim}{a}_0$ and $\underset{\sim}{b}_0$. Hence W can be expressed as a function of

$$I_1 , \quad I_2 , \quad I_3 , \quad I_8 , \quad I_9 = I_4 + I_6 , \quad I_{10} = I_4 I_6 ,$$

$$I_{11} = I_5 + I_7 , \quad I_{12} = I_5 I_7 \quad \text{and} \quad \cos^2 2\Phi . \tag{73}$$

However, it can be shown that I_{12} can be expressed in terms of the other invariants, and so W can be expressed as a function of the seven invariants

$$I_1, \quad I_2, \quad I_3, \quad I_8, \quad I_9, \quad I_{10}, \quad I_{11}$$

$$\text{and} \quad \cos^2 2\Phi . \tag{74}$$

Alternatively, W can be expressed in terms of the mutually orthogonal unit vectors which bisect $\underset{\sim}{a}_0$ and $\underset{\sim}{b}_0$, but because material line elements which bisect fibre directions in the reference configuration do not, in general, bisect fibre directions in the deformed configuration, it is not usually advantageous to do this.

When W is expressed as a function of the set (74), we obtain from (58)

$$\begin{aligned}\underset{\sim}{\sigma} = 2I_3^{-\frac{1}{2}}\{ &(I_2W_2+I_3W_3)\underset{\sim}{I} + W_1\underset{\sim}{B} - W_2\underset{\sim}{B}^{-1} + (I_4W_9+I_{10}W_{10})\underset{\sim}{a}\otimes\underset{\sim}{a} \\ &+ (I_6W_9+I_{10}W_{10})\underset{\sim}{b}\otimes\underset{\sim}{b} + \tfrac{1}{2}I_{10}^{\frac{1}{2}}W_8(\underset{\sim}{a}\otimes\underset{\sim}{b}+\underset{\sim}{b}\otimes\underset{\sim}{a}) \\ &+ I_4W_{11}(\underset{\sim}{a}\otimes\underset{\sim}{B}.\underset{\sim}{a}+\underset{\sim}{a}.\underset{\sim}{B}\otimes\underset{\sim}{a}) + I_6W_{11}(\underset{\sim}{b}\otimes\underset{\sim}{B}.\underset{\sim}{b}+\underset{\sim}{b}.\underset{\sim}{B}\otimes\underset{\sim}{b})\} . \end{aligned} \tag{75}$$

If the material is incompressible and inextensible in the two fibre directions, then $I_3 = I_4 = I_6 = 1$, and hence $I_9 = 2$, $I_{10} = 1$, and (75) is replaced by

$$\begin{aligned}\underset{\sim}{\sigma} = 2\{&W_1\underset{\sim}{B} - W_2\underset{\sim}{B}^{-1} + \tfrac{1}{2}W_8(\underset{\sim}{a}\otimes\underset{\sim}{b}+\underset{\sim}{b}\otimes\underset{\sim}{a}) + W_{11}(\underset{\sim}{a}\otimes\underset{\sim}{B}.\underset{\sim}{a}+\underset{\sim}{a}.\underset{\sim}{B}\otimes\underset{\sim}{a}+\underset{\sim}{b}\otimes\underset{\sim}{B}.\underset{\sim}{b}+\underset{\sim}{b}.\underset{\sim}{B}\otimes\underset{\sim}{b})\} \\ &- p\underset{\sim}{I} + T_a\underset{\sim}{a}\otimes\underset{\sim}{a} + T_b\underset{\sim}{b}\otimes\underset{\sim}{b} , \end{aligned} \tag{76}$$

where again the last three terms represent the reaction stress.

4. PLASTICITY THEORY FOR FIBRE-REINFORCED MATERIAL

4.1 Yield functions for one family of fibres

We now consider that the material has plastic response. We follow a standard formulation of plasticity theory, beginning with the yield function. The applications we have in mind are to fibre-reinforced composites with metal matrices, but the theory is not limited to any particular type of material.

We postulate a yield function $f(\sigma_{ij})$ such that in admissible stress states $f \leqslant 0$, with $f = 0$ when plastic deformation is taking place. For strain-hardening material some slight modifications are needed, which we take up later.

If the plastic material is isotropic then it is well known that f can be expressed as a function of the stress invariants $\mathrm{tr}\,\underset{\sim}{\sigma}$, $\mathrm{tr}\,\underset{\sim}{\sigma}^2$ and $\mathrm{tr}\,\underset{\sim}{\sigma}^3$. In isotropic metal plasticity it is observed experimentally that for many materials yielding is effectively independent of a superposed hydrostatic pressure. This is incorporated in the theory by restricting f to depend on the deviatoric stress $\underset{\sim}{s}$,

$$s_{ij} = \sigma_{ij} - \tfrac{1}{3}\sigma_{kk}\delta_{ij}, \qquad \underset{\sim}{s} = \underset{\sim}{\sigma} - \tfrac{1}{3}\underset{\sim}{I}\,\mathrm{tr}\,\underset{\sim}{\sigma}. \tag{77}$$

Then $\mathrm{tr}\,\underset{\sim}{s} = 0$ and f can be expressed as a function of $\mathrm{tr}\,s^2$ and $\mathrm{tr}\,\underset{\sim}{s}^3$. We note that $\underset{\sim}{s}$ is the extra-stress for an incompressible material.

For anisotropic materials f is a function of σ_{ij} (or s_{ij}) which is invariant under the appropriate transformation group.

For a fibre-reinforced material the yield properties will depend on the orientation of the fibres, so we propose f to be a function of σ_{ij} and a_i, or, since the sense of $\underset{\sim}{a}$ has no significance, of σ_{ij} and $a_i a_j$.

For a fibre-reinforced metal we expect yielding to remain independent of superposed hydrostatic stress. Also for a metal reinforced by inextensible fibres it is reasonable to expect that yielding is not affected by a superposed tension in the fibre direction, since such a tension produces no stress in the matrix. These conditions can be incorporated by assuming f to depend on $\underset{\sim}{\sigma}$ only through the extra-stress $\underset{\sim}{s}$, where

$$\underset{\sim}{\sigma} = \underset{\sim}{r} + \underset{\sim}{s}, \qquad \underset{\sim}{r} = -p\underset{\sim}{I} + T\underset{\sim}{a} \otimes \underset{\sim}{a}. \tag{78}$$

Here $\underset{\sim}{r}$ is the reaction stress and the indeterminacy in $\underset{\sim}{s}$ is removed by imposing the conditions

$$\mathrm{tr}\,\underset{\sim}{s} = 0, \qquad \underset{\sim}{a}.\underset{\sim}{s}.\underset{\sim}{a} = 0. \tag{79}$$

Then it follows from (78) and (79) that

$$\mathrm{tr}\,\underset{\sim}{\sigma} = -3p + T, \qquad \underset{\sim}{a}.\underset{\sim}{\sigma}.\underset{\sim}{a} = -p + T,$$

and it follows by eliminating p and T from (78) that

$$\underset{\sim}{s} = \underset{\sim}{\sigma} - \tfrac{1}{2}(\mathrm{tr}\,\underset{\sim}{\sigma} - \underset{\sim}{a}.\underset{\sim}{\sigma}.\underset{\sim}{a})\underset{\sim}{I} + \tfrac{1}{2}(\mathrm{tr}\,\underset{\sim}{\sigma} - 3\underset{\sim}{a}.\underset{\sim}{\sigma}.\underset{\sim}{a})\underset{\sim}{a}\otimes\underset{\sim}{a}\,. \tag{80}$$

Now f must be invariant under rotations of the stress field, with the fibres moving with the stress field. Hence, for any orthogonal tensor $\underset{\sim}{Q}$, if

$$\bar{\underset{\sim}{s}} = \underset{\sim}{Q}.\underset{\sim}{s}.\underset{\sim}{Q}^T, \qquad \bar{\underset{\sim}{a}} = \underset{\sim}{Q}.\underset{\sim}{a}$$

then we require

$$f(\underset{\sim}{s}, \underset{\sim}{a}\otimes\underset{\sim}{a}) = f(\bar{\underset{\sim}{s}}, \bar{\underset{\sim}{a}}\otimes\bar{\underset{\sim}{a}})\,.$$

Hence f is an isotropic invariant of $\underset{\sim}{s}$ and $\underset{\sim}{a}\otimes\underset{\sim}{a}$. By standard results in invariant theory [1], and taking into account (79) and (6), it follows that f can be expressed as a function of

$$J_1 = \mathrm{tr}\,\underset{\sim}{s}^2, \qquad J_2 = \underset{\sim}{a}.\underset{\sim}{s}^2.\underset{\sim}{a}, \qquad J_3 = \mathrm{tr}\,\underset{\sim}{s}^3\,. \tag{81}$$

In the solution of problems, even in isotropic plasticity theory, it is usually necessary to assume some special form for the yield function, the most commonly adopted yield functions being those of von Mises and Tresca. The natural approach for a fibre-reinforced material is to try to generalise these. Von Mises' yield function is the most general isotropic yield function which is quadratic in the stress components. The most general function of the invariants (81) which is quadratic in the stress is

$$f = \frac{1}{2k_T^2} J_1 + \left(\frac{1}{k_L^2} - \frac{1}{k_T^2}\right) J_2 - 1\,. \tag{82}$$

The coefficients of J_1 and J_2 are written in this way because it can then easily be shown that k_T and k_L can be identified with the shear yield stress on the planes containing the fibres for shear in the directions transverse to and parallel to the fibres respectively.

If $k_T = k_L$ (or, more generally, if f depends on J_1 and J_3 only), the fibre orientation does not enter into the yield condition, and the material behaves as a *constrained isotropic material*. In such a material the extra-stress response is isotropic. However, the experimental evidence is that $k_T \neq k_L$, except perhaps for low fibre densities.

A yield function analogous to Tresca's, which is essentially due to Lance and Robinson [8] is

$$f = \begin{cases} \dfrac{(\frac{1}{2}J_1 - J_2)^{\frac{1}{2}}}{k_T} - 1, & \text{for } J_2 < k_L^2, \\[2ex] \dfrac{J_2^{\frac{1}{2}}}{k_L} - 1, & \text{for } (\frac{1}{2}J_1 - J_2) < k_T^2, \end{cases} \tag{83}$$

This effectively states that yield occurs when the component in either the transverse direction or the longitudinal direction of the shear stress on planes containing the fibres reaches a critical value.

As an example, consider uniaxial tension in the x_1-direction of a rectangular block of material reinforced by straight parallel fibres in the (x_1, x_2) planes and inclined at an angle ϕ to the x_1-axis. Thus

$$\sigma_{22} = \sigma_{33} = \sigma_{23} = \sigma_{31} = \sigma_{12} = 0, \qquad \underset{\sim}{a} = (\cos\phi, \sin\phi, 0),$$

which gives

$$J_1 = \tfrac{1}{2}\sigma_{11}^2 \sin^2\phi(1 + 3\cos^2\phi), \qquad J_2 = \sigma_{11}^2 \sin^2\phi \cos^2\phi.$$

On substituting these into (82) we find that $f = 0$ when

$$\sigma_{11} = \frac{\pm 2k_L k_T}{\sin\phi \cos\phi\,(k_L^2 \tan^2\phi + 4k_T^2)^{\frac{1}{2}}}. \tag{84}$$

Alternatively, the yield condition (83) gives in this case

$$\sigma_{11} = \begin{cases} \pm k_L/\sin\phi\cos\phi, & |\tan\phi| < 2k_T/k_L, \\[2ex] \pm 2k_T/\sin^2\phi, & |\tan\phi| > 2k_T/k_L. \end{cases} \tag{85}$$

Both (84) and (85) give very good fits to experimental data with appropriate choices of k_T and k_L.

4.2 Yield functions for two families of fibres

Similar arguments can be used in the case of two families of fibres. In this case the extra-stress $\underset{\sim}{s}$ satisfies the conditions

$$\operatorname{tr}\underset{\sim}{s} = 0\,, \qquad \underset{\sim}{a}.\underset{\sim}{s}.\underset{\sim}{a} = 0\,, \qquad \underset{\sim}{b}.\underset{\sim}{s}.\underset{\sim}{b} = 0\,, \tag{86}$$

and it follows, after some manipulation, that

$$\begin{aligned} \underset{\sim}{s} = \underset{\sim}{\sigma} + (1+3\cos^2 2\phi)^{-1}[&\{\underset{\sim}{a}.\underset{\sim}{\sigma}.\underset{\sim}{a}+\underset{\sim}{b}.\underset{\sim}{\sigma}.\underset{\sim}{b} - (1+\cos^2 2\phi)\operatorname{tr}\phi\}\underset{\sim}{I} \\ &+ \{\operatorname{tr}\underset{\sim}{\sigma} - (2\operatorname{cosec}^2 2\phi)\underset{\sim}{a}.\underset{\sim}{\sigma}.\underset{\sim}{a} - (\operatorname{cosec}^2 2\phi - 3\cot^2 2\phi)\underset{\sim}{b}.\underset{\sim}{\sigma}.\underset{\sim}{b}\}\underset{\sim}{a}\otimes\underset{\sim}{a} \\ &+ \{\operatorname{tr}\underset{\sim}{\sigma} - (2\operatorname{cosec}^2 2\phi)\underset{\sim}{b}.\underset{\sim}{\sigma}.\underset{\sim}{b} - (\operatorname{cosec}^2 2\phi - 3\cot^2 2\phi)\underset{\sim}{a}.\underset{\sim}{\sigma}.\underset{\sim}{a}\}\underset{\sim}{b}\otimes\underset{\sim}{b}]\,, \end{aligned} \tag{87}$$

where $\cos 2\phi = \underset{\sim}{a}.\underset{\sim}{b}$. Then it is assumed that f is a function of $\underset{\sim}{s}$, $\underset{\sim}{a}\otimes\underset{\sim}{a}$ and $\underset{\sim}{b}\otimes\underset{\sim}{b}$, and by arguments similar to those used above it follows that f is an isotropic invariant of these tensors. It then follows by standard results in invariant theory that f can be expressed as a function of

$$\begin{aligned} &J_1 = \operatorname{tr}\underset{\sim}{s}^2\,, \qquad J_2 = \underset{\sim}{a}.\underset{\sim}{s}^2.\underset{\sim}{a}\,, \qquad J_3 = \operatorname{tr}\underset{\sim}{s}^3\,, \qquad J_4 = \underset{\sim}{b}.\underset{\sim}{s}^2.\underset{\sim}{b}\,, \\ &J_5 = \cos 2\phi\, \underset{\sim}{a}.\underset{\sim}{s}.\underset{\sim}{b}\,, \qquad J_6 = \cos 2\phi\, \underset{\sim}{a}.\underset{\sim}{s}^2.\underset{\sim}{b} \qquad \text{and} \qquad \cos^2 2\phi\,. \end{aligned} \tag{88}$$

If the two families of fibres are mechanically equivalent, then f must be symmetrical with respect to interchanges of $\underset{\sim}{a}$ and $\underset{\sim}{b}$, and then dependence on J_2 and J_4 can be replaced by dependence on J_2+J_4 and J_2J_4. However J_2J_4 can be expressed in terms of J_1, J_3, J_5, J_6 and $\cos^2 2\phi$, and so it may be omitted. Thus in this case f becomes a function of J_1, J_2+J_4, J_3, J_5, J_6 and $\cos^2 2\phi$. The most general quadratic yield function is then (J_5^2 can be expressed in terms of other invariants and omitted)

$$f = \frac{J_2+J_4}{c_1^2} + \frac{J_1}{c_2^2} + \frac{J_6}{c_3^2} - 1\,, \tag{89}$$

where c_1, c_2 and c_3 have the dimensions of stress and are functions of $\cos^2 2\phi$.

As an example consider yielding of a rectangular block reinforced by two families of straight parallel fibres so that

$$\underset{\sim}{a} = (\cos\phi, \sin\phi, 0), \qquad \underset{\sim}{b} = (\cos\phi, -\sin\phi, 0). \tag{90}$$

By inserting (90) in (88), and expressing s_{ij} in terms of σ_{ij} using (87), it can be shown that in this configuration J_2+J_4, J_1 and J_6 are each linear combinations of σ_{13}^2, σ_{23}^2 and $\{(\sigma_{11}-\sigma_{33})\sin^2\phi - (\sigma_{22}-\sigma_{33})\cos^2\phi\}^2$. Hence for this configuration the yield function (89) can be expressed in the form

$$f = \frac{1}{Y^2}\{(\sigma_{11}-\sigma_{33})\sin^2\phi - (\sigma_{22}-\sigma_{33})\cos^2\phi\}^2 + \frac{\sigma_{13}^2}{k_1^2} + \frac{\sigma_{23}^2}{k_2^2} - 1, \tag{91}$$

where Y, k_1 and k_2 are functions of ϕ which, with some manipulation, can be related to c_1, c_2 and c_3. The parameters k_1 and k_2 can be interpreted as shear yield stresses for shear on planes x_3 = constant in the x_1 and x_2 directions respectively. If Y_1, Y_2 and Y_3 are defined by

$$Y = Y_1\sin^2\phi = Y_2\cos^2\phi = Y_3|\cos^2\phi - \sin^2\phi|, \tag{92}$$

then Y_1, Y_2 and Y_3 are yield stresses in uniaxial tension in the x_1, x_2 and x_3 directions, for this yield condition. Hence in principle k_1, k_2 and Y may be determined experimentally.

Suppose the block is subjected to simple tension P along an axis defined by the unit vector $(\cos\theta, \sin\theta, 0)$. Then

$$\sigma_{11} = \tfrac{1}{2}P(1+\cos 2\theta), \qquad \sigma_{22} = \tfrac{1}{2}P(1-\cos 2\theta), \qquad \sigma_{12} = \tfrac{1}{2}P\sin 2\theta,$$

$$\sigma_{13} = \sigma_{23} = \sigma_{33} = 0.$$

By substituting these in (91) we find that

$$P = \pm\frac{2Y}{\cos 2\theta - \cos 2\phi}. \tag{93}$$

The same yield stresses Y, k_1 and k_2 occur in problems of deformations of helically reinforced cylinders.

4.3 Flow rules

To complete a theory of plasticity, constitutive equations are required. One common procedure for formulating these is to assume that the yield function is also a plastic potential function, so that the components d^p_{ij} of the plastic strain-rate $\underset{\sim}{d}^p$ are given by

$$d^p_{ij} = \dot{\lambda}\partial f/\partial\sigma_{ij}\,, \tag{94}$$

where $\dot{\lambda}$ is a scalar multiplier (not a material constant). In a rigid-plastic theory $\underset{\sim}{d}^p$ is the total strain-rate $\underset{\sim}{d}$ with components d_{ij}, so that

$$d^p_{ij} = d_{ij} = \frac{1}{2}\left(\frac{\partial v_i}{\partial x_j} + \frac{\partial v_j}{\partial x_i}\right), \tag{95}$$

where v_i are components of velocity. There are various justifications for (94); for example, it may be deduced as a consequence of Drucker's stability postulate. The usual arguments leading to (94) remain valid in the presence of kinematic constraints. It is straightforward to calculate $\underset{\sim}{d}^p$ from (94) for the yield functions discussed in this chapter.

When f has the form (82), the flow rule (94) becomes

$$\underset{\sim}{d}^p = \dot{\lambda}\left\{\frac{1}{k_L^2}\underset{\sim}{s} + \left(\frac{1}{k_L^2} - \frac{1}{k_T^2}\right)(\underset{\sim}{a}\otimes\underset{\sim}{a}.\underset{\sim}{s}+\underset{\sim}{s}.\underset{\sim}{a}\otimes\underset{\sim}{a})\right\}, \tag{96}$$

and this can be expressed in terms of $\underset{\sim}{\sigma}$ by using (80). When f has the form (89), the flow rule becomes

$$\begin{aligned}\underset{\sim}{d}^p = \dot{\lambda}\{&(1+3\cos^2 2\phi)^{-1}J_5(4c_1^{-2} - 2c_3^{-2}\cos^2 2\phi)\underset{\sim}{I} + 2c_2^{-2}\underset{\sim}{s}\\ &+ (1+3\cos^2 2\phi)^{-1}J_5(6c_1^{-2}+c_3^{-2})(\underset{\sim}{a}\otimes\underset{\sim}{a}+\underset{\sim}{b}\otimes\underset{\sim}{b})\\ &+ c_1^{-2}(\underset{\sim}{a}\otimes\underset{\sim}{a}.\underset{\sim}{s}+\underset{\sim}{s}.\underset{\sim}{a}\otimes\underset{\sim}{a}+\underset{\sim}{b}\otimes\underset{\sim}{b}.\underset{\sim}{s}+\underset{\sim}{s}.\underset{\sim}{b}\otimes\underset{\sim}{b})\\ &+ \tfrac{1}{2}c_3^{-2}\cos 2\phi(\underset{\sim}{a}\otimes\underset{\sim}{b}.\underset{\sim}{s}+\underset{\sim}{s}.\underset{\sim}{b}\otimes\underset{\sim}{a}+\underset{\sim}{b}\otimes\underset{\sim}{a}.\underset{\sim}{s}+\underset{\sim}{s}.\underset{\sim}{a}\otimes\underset{\sim}{b})\}\,,\end{aligned} \tag{97}$$

where $\underset{\sim}{s}$ is now given by (87).

The $\underset{\sim}{d}^p$ obtained as above automatically satisfy the constraints of plastic incompressibility and fibre inextensibility.

4.4 Hardening rules

For a perfectly plastic (i.e. non-hardening) material the yield stresses, such as k_L and k_T in (82), are constants. We now consider the possibility that the material is work- or strain-hardening.

In general the hardening properties of a highly anisotropic plastic solid will be complicated, as even in the simplest cases (e.g. (82)) several parameters are required to describe the current yield surface. A major simplication results if it is assumed that the current state of hardening of the material can be described by a single parameter, which we take to be the plastic work W_p, defined by

$$\dot{W}_p = \sigma_{ij} d^p_{ij} = s_{ij} d^p_{ij} . \tag{98}$$

This point of view is doubtless an oversimplification of the real situation, but is plausible in some circumstances - for example, if we consider a fibre-reinforced composite with an isotropically-hardening metal matrix. Then it is reasonable to assume that the state of hardening of the composite is controlled by the state of hardening of the matrix, which in turn, for isotropic hardening, depends on a single parameter such as the plastic work.

We therefore assume that the yield condition can be expressed in the form

$$g(\sigma_{ij}) = k(W_p) , \tag{99}$$

where k has the dimensions of stress. We further assume that $g(\sigma_{ij}) = k$ is a convex surface in σ_{ij} space and that $\{g(\sigma_{ij})\}^2$ is a homogeneous function of degree two in σ_{ij} (this is not quite the same as assuming g is homogeneous of degree one; this formulation avoids ambiguities which arise when taking square roots).

By the flow rule (94)

$$d^p_{ij} = \dot{\varepsilon}\partial g/\partial\sigma_{ij} , \tag{100}$$

and $\dot{\varepsilon}$ has dimensions $(\text{time})^{-1}$. Then

$$\dot{W}_p = \dot{\varepsilon}\sigma_{ij}\partial g/\partial\sigma_{ij} = \dot{\varepsilon}g(\sigma_{ij}) = \dot{\varepsilon}k(W_p) , \tag{101}$$

by Euler's theorem for homogeneous functions. Equation (101) establishes a correspondence between W_p and ε; if k is an increasing function of W_p

this correspondence is one-to-one. Hence k may be regarded as a function of $\dot{\varepsilon}$ rather than of W_p. Also $\dot{\varepsilon}$ has the dimensions of a strain-rate, and so may be regarded as an equivalent strain-rate. To obtain an explicit expression for $\dot{\varepsilon}$ it is necessary to solve (100) for σ_{ij} in terms of d^p_{ij} and $\dot{\varepsilon}$. Then substituting for σ_{ij} in the yield condition gives $\dot{\varepsilon}$ as a function of d^p_{ij}.

In practice it may be difficult to express the yield condition in the form (99). For example this is the case with the yield function (82) when k_T and k_L are regarded as functions of W_p or ε. If we simplify further and suppose that the ratio $k_T/k_L = \alpha$ remains constant (this assumption can also be made plausible), then the yield condition corresponding to (82) takes the form

$$\{\tfrac{1}{2}J_1+(\alpha^2-1)J_2\}^{\frac{1}{2}} = k_T(\varepsilon)\,, \tag{102}$$

which is of the form (99). Then $\dot{\varepsilon}$ can be related to the stress-rate by

$$\frac{dk_T}{d\varepsilon}\dot{\varepsilon} = \frac{\tfrac{1}{2}\dot{J}_1+(\alpha^2-1)\dot{J}_2}{2\{\tfrac{1}{2}J_1+(\alpha^2-1)J_2\}^{\frac{1}{2}}}\,. \tag{103}$$

Also from (100) and (102) (and equivalently to (96) when $k_T = \alpha k_L$)

$$d^p_{ij} = \frac{\dot{\varepsilon}}{2k_T}\{s_{ij}+(\alpha^2-1)(a_i a_k s_{kj}+a_j a_k s_{ki})\} \tag{104}$$

and from (102) and (104) it follows after a little manipulation that

$$\dot{\varepsilon}^2 = 2\{d^p_{ij}d^p_{ij}+2(\alpha^{-2}-1)a_i a_j d^p_{ik}d^p_{jk}\}\,. \tag{105}$$

Some simpler problems for strain-hardening materials will be considered in Chapter IX in connection with dynamic problems for beams and plates.

4.5 Small elastic-plastic deformations

So far we have considered only rigid-plastic materials. In constructing an elastic-plastic theory we restrict attention to problems of small deformations. Then the usual procedure, which we follow, is to assume that the strain-rate tensor $\underset{\sim}{d}$ is the sum of an elastic strain-rate

$\underset{\sim}{d}^e$ and a plastic strain-rate $\underset{\sim}{d}^p$, so that

$$\underset{\sim}{d} = \underset{\sim}{d}^e + \underset{\sim}{d}^p, \qquad d_{ij} = d_{ij}^e + d_{ij}^p. \tag{106}$$

We consider that the elastic strain-rate depends linearly on the stress-rate, and so the constitutive equations for the elastic strain-rate are analogous to the linear elastic stress-strain relations, described earlier, and require no further discussion.

For a composite comprising a ductile matrix reinforced with elastic fibres, it is plausible that only the elastic strain contributes to the volume change and to the extensions in the fibre directions. Consequently the plastic strain involves no volume change or fibre extensions. If the flow rule (94) is adopted, this means that the yield function f must be a function of the invariants (81) for a single family of fibres and (88) for two families of fibres. Thus the yield function takes the same form as in the rigid-plastic theory, and $\underset{\sim}{d}^p$ is given by the flow rule (94), so that the results given above for the rigid-plastic theory apply provided that $\underset{\sim}{d}^p$ is interpreted as the plastic part of the strain-rate.

REFERENCES

[1] SPENCER, A.J.M., Theory of invariants, in *Continuum Physics, Vol.1*, Eringen, A.C., Ed., Academic Press, New York (1971) 239-253

[2] SPENCER, A.J.M., The formulation of constitutive equations for anisotropic solids, in *Comportement mécanique des solides anisotropes (Colloques internationaux du C.N.R.S. No.295)*, Boehler, J-P. and Sawczuk, A., Eds., Editions Scientifiques du C.N.R.S., Paris (1982) 3-26

[3] MARKHAM, M.F., *Composites* 1 (1970) 145-149

[4] SPENCER, A.J.M., *Continuum Mechanics*, Longman, London, 1980

[5] ERICKSEN, J.E. and RIVLIN, R.S., *J. Rat. Mech. Anal.* 3 (1954) 281-301

[6] SMITH, G.F. and RIVLIN, R.S., *Trans. Amer. Math. Soc.* 88 (1958) 175-193

[7] GREEN, A.E. and ADKINS, J.E., *Large Elastic Deformations*, Clarendon Press, Oxford, 1960

[8] LANCE, R.H. and ROBINSON, D.N., *J. Mech. Phys. Solids* 19 (1971) 49-60

Additional bibliography

PIPKIN, A.C., Finite deformations of ideal fiber-reinforced composites, in *Composite Materials, Vol.2*, Sendeckyj, G.P., Ed., Academic Press, New York (1973) 251-308

PIPKIN, A.C., *Advances in Applied Mechanics* 19 (1979) 1-51

ROGERS, T.G., Anisotropic elastic and plastic materials, in *Continuum Mechanics Aspects of Geodynamics and Rock Fracture Mechanics*, Thoft-Christensen, P., Ed., Reidel, Dordrecht (1975) 177-200

ROGERS, T.G., Finite deformations of strongly anisotropic materials, in *Theoretical Rheology*, Hutton, J.F., Pearson, J.R.A. and Walters, K., Eds., Applied Science Publishers, London (1975) 141-168

SPENCER, A.J.M., *Deformations of Fibre-reinforced Materials*, Clarendon Press, Oxford, 1972

SPENCER, A.J.M., The formulation of constitutive equations in continuum models of fibre-reinforced composites, in *Proceedings of the Third Symposium on Continuum Models of Discrete Systems*, Kröner, E. and Anthony, K-H., Eds., University of Waterloo Press (1980) 479-489

SPENCER, A.J.M., Continuum models of fibre-reinforced materials, in *Proceedings of the International Symposium on the Mechanical Behaviour of Structured Media*, Selvadurai, A.P.S., Ed., Elsevier, Amsterdam (1981) 1-26

II
FINITE DEFORMATION AND STRESS IN IDEAL FIBRE-REINFORCED MATERIALS

T. G. ROGERS

Department of Theoretical Mechanics
University of Nottingham
Nottingham, NG7 2RD
England

1. THEORY OF IDEAL FIBRE-REINFORCED MATERIALS (IFRM)

1.1 Introduction

The constitutive theory for strongly anisotropic materials is treated elsewhere (Chapter I) in this book. These materials are characterised by their physical property of having 'strong' directions, for each of which the extensional modulus is much greater than the shear moduli associated with that direction. This property is particularly true of many of the man-made fibre-reinforced composites which are now coming into widespread use; in these a relatively weak, isotropic matrix with certain desirable properties (such as lightness or ductility) is strengthened throughout in one or more particular directions by introducing strong reinforcing fibres in those directions. Macroscopically, these composite materials will exhibit mechanical properties which are transversely isotropic if reinforced by one family of fibres, for example, or orthotropic if reinforced by two families (refer Chapter I). This will be so even if all the constituents are isotropic themselves. Moreover, the response of these composites is not just anisotropic - it is usually *highly* anisotropic, so that isotropic theory would not provide even a rough approximation to their behaviour under most types of loading conditions.

The *general* constitutive equations derived in Chapter I for anisotropic elastic, viscoelastic or plastic behaviours are usually too complicated for solving any significant boundary-value problems. This is particularly true when the deformation is finite or if the reinforcement initially lies in curved lines. Nevertheless, a design engineer or stress analyst working with such materials still requires a theory which incorporates the highly anisotropic properties whilst at the same time being sufficiently simple to cope with fairly complex geometries and stress situations.

These considerations have led to the development of the theory of ideal fibre-reinforced materials (IFRM), which is the subject of much of this book. The characteristic properties of a strongly anisotropic material are represented by an idealised model in which

(i) there is no extension or contraction in a 'strong' direction,

(ii) the reinforcement is continuously distributed throughout the body and deforms with the material, and

(iii) the material is incompressible.

The first assumption recognises that if the material has *any* alternative to extending (or contracting) in a 'strong' direction, then it will take it, usually by shearing along or transverse to that direction. The second assumption is based on the fact that in a real fibre-reinforced material the actual fibres are convected with the adjoining matrix as the composite deforms. Assuming incompressibility merely follows the usual practice in theories of finite elasticity and plasticity, and is not related to whether or not the material has strength in any particular direction; just as for the conventional theories, we expect the assumption to be reasonable whenever large deformations are involved (or for the many situations involving small deformations in which the change of volume is negligible).

We note that all these assumptions defining IFRM are kinematical ones. Although they have implications for the stress response (refer Chapter I), no additional assumption is made regarding the constitutive equation relating the stress and deformation of the material. We shall show that in fact much of the IFRM theory is independent of the shearing stress response of the materials [1,2]. Thus, although the first theory of this sort concerned finite *elastic* behaviour [3], the first significant development of the theory came in the context of *plastic* deformations [4]. The theory has subsequently been extended to include all types of material response, and to give solutions to a number of types of boundary-value problems involving elastic, plastic or viscoelastic behaviour. It can treat large deformations as well as small ones, and the reinforcements can be in curved lines as well as straight.

This section of the course considers the relevant finite deformation theory. A long review of this theory has been given by Pipkin [5] and shorter reviews by Rogers [6,7] and Pipkin [8]. Spencer's book [2] contains most of the basic theory discussed here, and in much more detail.

We first present a few general results and then particularise to studies of boundary value problems involving either plane strain or axisymmetric deformations. By virtue of their introducing yet another kinematical condition, we should expect that both these two-dimensional theories would provide the simplest analysis, and indeed much of the analysis is then reduced to kinematics.

The plane strain theory is particularly simple, whatever the material behaviour and is described in Section 2. Explicit solutions can be obtained for many problems which even in conventional linear elasticity are extremely cumbersome or even impossible to solve. Displacement boundary value problems can be solved purely kinematically, without reference to the stress response nor even to the equilibrium equations. In traction or mixed boundary-value problems the solution requires a constitutive assumption to be made, but even here the stress analysis can usually be done by very elementary methods.

The theory also shows up clearly how different the response of strongly anisotropic bodies can be as compared with that of isotropic bodies. For example, end effects are not localised to one or two typical diameters' distance from the boundary, but can possibly extend the entire length of the body; this implies that the conventional use of St. Venant's principle must be radically amended if it is to be applicable to problems involving strongly anisotropic materials. Another remarkable result of the theory is the possibility of having *sheets* of singular stress both inside and on the outer surfaces of a body [1]; this contrasts with the conventional *points* of singular stress sometimes predicted in solutions in isotropic linear elasticity.

These features, and the ease of IFRM solutions, are demonstrated in Section 3, where we analyse a few simple but illustrative problems.

Axisymmetric deformations without twist are treated in Section 4. The relevant theory is not so simple as that of plane strain, and is relatively unexploited, this is reflected in the brevity of the section.

We end this introductory section by reminding the reader that whereas the motivation for the theory was provided by the new fibre-reinforced composites, the theory itself is appropriate for *any* material which exhibits strong anisotropy. Nevertheless, even for such cases we still adopt the terms 'fibre' and 'fibre-direction' to describe the preferred, 'strong' directions. This not only emphasises the highly directional properties of such a material, but should also enable readers to visualise and understand more easily the predictions of the theory.

1.2 Kinematics

We adopt the same notation as in Chapter I, except that we now use $\underset{\sim}{A}(\underset{\sim}{X})$, $\underset{\sim}{B}(\underset{\sim}{X})$, ... to denote the initial fibre-directions at an element at $\underset{\sim}{X}$.

The analytical implications of the kinematical constraints *(i)* - *(iii)* are straightforward. Fibre-inextensibility implies

$$|\underset{\sim}{A}(\underset{\sim}{X})| = |\underset{\sim}{a}(\underset{\sim}{X},0)| = |\underset{\sim}{a}(\underset{\sim}{X},t)| = 1\,, \tag{1}$$

where, for convenience, we make $\underset{\sim}{A}$ and $\underset{\sim}{a}$ unit vectors. From the second property we deduce that

$$a_i = A_R \partial x_i/\partial X_R = A_R F_{iR} , \tag{2}$$

so that, with (1), we obtain

$$A_R A_S C_{RS} = A_R A_S F_{iR} F_{iS} = 1 , \tag{3}$$

with similar results for all other fibre-directions $\underset{\sim}{B}, \ldots$ at $\underset{\sim}{X}$.

Incompressibility implies (see Chapter I)

$$J \equiv \det \underset{\sim}{F} = \det \underset{\sim}{C} = 1 . \tag{4}$$

Furthermore, this constraint also implies that fibre flux is conserved throughout a deformation, whether the material is inextensible or not (for proof, see [1], [2] or [9]). Thus

$$\frac{\partial a_i}{\partial x_i} = \frac{\partial A_R}{\partial X_R} \quad \text{or} \quad \underset{\sim}{\nabla} . \underset{\sim}{a} = \underset{\sim}{\nabla}_0 . \underset{\sim}{A} . \tag{5}$$

With fibre inextensibility, both $\underset{\sim}{a}$ and $\underset{\sim}{A}$ are fields of unit vectors, and these divergences are then related to the curvatures of curves orthogonal to the fibres.

Cases in which the initial fibre-flux vanishes are particularly important because these correspond to the usual practical situation in which the reinforcing cords are packed as closely together as possible, with a negligible number ending in the interior of the composite. In such cases, (5) shows that the fibre-flux is *always* zero.

The result (5) also has implications for bent fibres. It is kinematically admissible for fibres to change direction discontinuously as they pass through an interior surface. If they do, changing direction from $\underset{\sim}{a}^-$ to $\underset{\sim}{a}^+$ as they pass through a surface element $\underset{\sim}{n}$ dS, then continuity of the fibre flux $\underset{\sim}{a}.\underset{\sim}{n}$ dS implies that

$$\underset{\sim}{n}.\underset{\sim}{a}^- = \underset{\sim}{n}.\underset{\sim}{a}^+ . \tag{6}$$

In two dimensions, this implies that the line of discontinuity bisects the angle between the fibre-directions on each side of it.

1.3 Stress and equilibrium

As described in Chapter I, it is convenient to divide the stress $\underset{\sim}{\sigma}$

into two parts:

$$\underset{\sim}{\sigma} = \underset{\sim}{s} + \underset{\sim}{r} \qquad \text{or} \qquad \sigma_{ij} = s_{ij} + r_{ij}\,, \tag{7}$$

where $\underset{\sim}{r}$ represents the reaction stress and $\underset{\sim}{s}$ is the extra-stress. The extra stress $\underset{\sim}{s}$ depends on the particular material being considered and is related to the deformation through a constitutive equation. The remainder of the stress is the reaction to the kinematical constraints and is arbitrary in the sense that it is independent of the deformation. It is determined by the total stress having to satisfy the equilibrium equations and appropriate traction boundary conditions.

Thus for an ideal fibre-reinforced material with a *single* family of fibres the reaction stress is of the form

$$\underset{\sim}{r} = T\underset{\sim}{a}\otimes\underset{\sim}{a} - p(\underset{\sim}{I}-\underset{\sim}{a}\otimes\underset{\sim}{a})$$

so that (7) becomes

$$\sigma_{ij} = Ta_i a_j - p(\delta_{ij}-a_i a_j) + s_{ij}\,. \tag{8}$$

Without loss of generality, the terms s_{ii} and $a_i a_j s_{ij}$ can be absorbed into the arbitrary functions of position p and T so that we obtain the convenient relations

$$s_{ii} = 0\,, \qquad a_i a_j s_{ij} = 0\,. \tag{9}$$

As described in Chapter I, the term $T\underset{\sim}{a}\otimes\underset{\sim}{a}$ represents a tensile stress of magnitude T in the fibre-direction $\underset{\sim}{a}$; the isotropic pressure $-p\underset{\sim}{I}$ is the reaction to the constraint of incompressibility. It is convenient to subtract from the latter stress its component in the fibre-direction so that T stands for the *total* tensile stress on surfaces normal to the $\underset{\sim}{a}$-direction.

In this entire chapter, we consider only problems of equilibrium and, for convenience, we neglect body forces. Dynamical effects are treated elsewhere (Chapters VIII and IX). Then the stresses satisfy the equations of equilibrium:

$$\frac{\partial\sigma_{ij}}{\partial x_j} = 0\,.$$

By substitution of (8) we obtain

$$\frac{\partial p}{\partial x_i} - a_i a_j \frac{\partial T'}{\partial x_j} - \left(a_i \frac{\partial a_j}{\partial x_j} + a_j \frac{\partial a_i}{\partial x_j}\right) T' = \frac{\partial s_{ij}}{\partial x_j}, \tag{10}$$

where

$$T' = T + p. \tag{11}$$

If a kinematically admissible deformation has been found, then the a_i are known and the s_{ij} determined from the constitutive equations and (9). Then (10) become three equations for the two functions p and T' (and hence T). In general these equations will not have a solution, so in fact they will impose a further restriction that the kinematically admissible solution must satisfy.

A deformation for which (10) can be satisfied by appropriate p and T' is called 'statically admissible'. In order to determine such deformations it is usually necessary first to express s_{ij} in terms of the deformation, using the constitutive equations.

The cases of plane strain and axial symmetry provide enough symmetry to reduce (10) to two independent equations; hence problems with either of these properties can furnish solutions for p and T', whatever the form of the constitutive equation. It will of course be necessary that appropriate surface tractions can be applied to support the deformation in the relevant boundary-value problem. In such cases we then have the powerful result that *for every kinematically admissible deformation there is a statically admissible stress field compatible with it*

The physical significance of a 'strong' direction suggests that the fibres themselves might conveniently be used as reference lines. Then for the one-family IFRM an obvious choice for two such intrinsic coordinates would be arc-lengths along the fibre-direction $\underset{\sim}{a}$ and along trajectories in directions $\underset{\sim}{n}$ normal to $\underset{\sim}{a}$. We note that, unlike fibres, such normal lines are *not* material lines in general; a set of particles lying along a normal line in the undeformed state will ordinarily *not* lie along a normal line when the body is deformed. In these particular directions the equilibrium equations (10) take simpler forms; thus the component of (10) in the fibre-direction $\underset{\sim}{a}$ is

$$\frac{\partial T}{\partial a} + (T+p)\frac{\partial a_i}{\partial x_i} = -a_i \frac{\partial s_{ij}}{\partial x_j} \tag{12}$$

and the component in a direction $\underset{\sim}{n}$ normal to $\underset{\sim}{a}$ is

$$\frac{\partial p}{\partial n} - (T+p)n_i a_j \frac{\partial a_i}{\partial x_j} = n_i \frac{\partial s_{ij}}{\partial x_j}, \tag{13}$$

where $\partial/\partial a$ and $\partial/\partial n$ represent differentiation with respect to arc-length along the fibres ($\underset{\sim}{a}$-lines) and the normal lines ($\underset{\sim}{n}$-lines).

In the important special case of close-packed fibres ($\nabla.\underset{\sim}{a} = 0$), equation (12) simplifies to give T directly without p needing to be known; then (13) becomes an ordinary differential equation governing the variation of p along the $\underset{\sim}{n}$-lines.

For materials with two fibre-families with directions $\underset{\sim}{a}$ and $\underset{\sim}{b}$, the equilibrium equations furnish three equations for the functions p, T_a and T_b (see Chapter I). Then, in general, in this case any kinematically admissible deformation is also statically admissible. This is true irrespective of the form of the constitutive equation, although the actual stress field associated with the deformation does depend on a particular material behaviour through the constitutive equation.

In the remainder of this chapter, we restrict attention to the case of a single family of reinforcing fibres. Theory involving two or more families of fibres is considered in Spencer's book [2], and the particular, but well-developed, case of cylinders with two helical families of fibres is described in Chapter V.

1.4 Stress discontinuities and singular stresses

We first consider the surfaces of discontinuity, considered previously in the chapter, across which the fibre-direction changes discontinuously. For such surfaces the stress can also change discontinuously, but in such a way that equilibrium is maintained everywhere; thus the normal stress and tangential shear stress, defined by

$$\sigma_{\nu\nu} = \sigma_{ij}\nu_i\nu_j \qquad \text{and} \qquad \sigma_{\nu\tau} = \sigma_{ij}\nu_i\tau_j, \tag{14}$$

must be continuous across the surface with unit outward normal $\underset{\sim}{\nu}$. With (8) these continuity conditions yield expressions for T and p on one side of the surface in terms of their values on the other side.

Such considerations are no different from similar situations in conventional theories, such as occur when two dissimilar materials are bonded together. However, in IFRM theory there is another kind of stress discontinuity which is a direct consequence of the basic assumptions of the theory. *The shearing stress can be discontinuous across a fibre surface.*

This possibility can be understood by considering a simple problem [8] (Figure 1). A slab, uniformly reinforced by straight fibres, is finitely deformed by pulling on the central sheet of fibres while the two boundaries are held parallel to the fibres. This produces two shearing deformations, one on each side of the sheet. The shearing forces on the two sides of the sheet both act in the same direction, so that the fibre-sheet itself must carry a finite force. In IFRM theory, this sheet has zero thickness, so in-plane tensile stress in it is infinite, throughout the length of the slab. The IFRM theory permits this because the fibres, being inextensible, can support any stress however large. The analysis in this case would show that the fibre-tension T involves a term $F(x)\delta(y)$, where $\delta(y)$ is a Dirac delta function and $F(x)$ is the finite force carried by the fibre y = 0.

Stress concentration layers of this kind can also occur in some 'normal' surfaces, as will be demonstrated for plane deformations (next

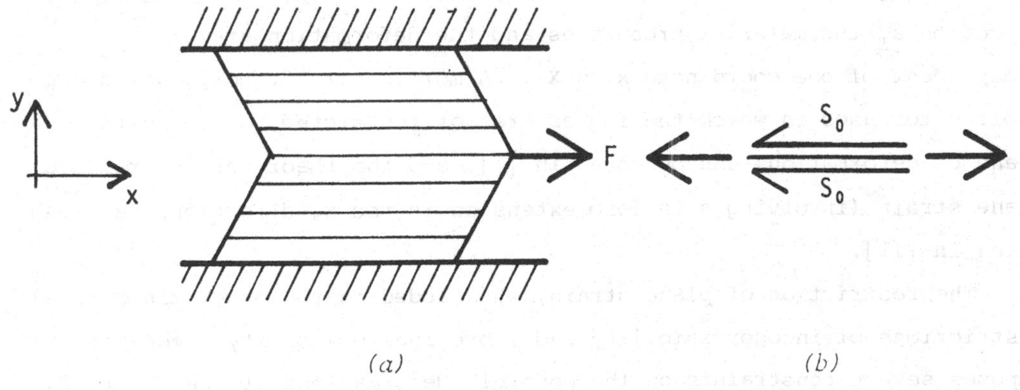

Figure 1. Fibre carrying a finite load

section) and elsewhere (Chapter V). In general they cannot occur in any other directions since usually an infinite tension produces infinite deformation.

Singular fibre layers are predicted frequently by the theory. They occur not only when concentrated loads are applied to the boundary but also whenever a free *flexed* surface of a body contains the fibres and does not intersect them. The interpretation and implications (for fracture, for example) of these stress concentration layers are well understood in the context of infinitesimal deformations, and is discussed in detail elsewhere (Chapters IV and VI). It is expected that a similar interpretation - thin layers of very high in-surface stresses - also applies to singular surfaces which occur in finitely deformed IFRM bodies. This expectation has been confirmed recently by Spencer [10] for two particular problems for slightly extensible transversely isotropic finite elastic materials.

2. PLANE FINITE DEFORMATIONS

2.1 Introduction

In this section and the next we describe the important special case of finite deformations of bodies in plane strain. The treatment here is based on [1], and considers only those problems in which the single fibre-direction $\underset{\sim}{a}$, the material properties and the deformation are all independent of one coordinate $x_3 = X_3$. A more general theory, which applies to cases in which the fibres are not restricted to lie in the plane of deformation, can be found in [2], and the theory of generalised plane strain (involving a uniform extension in the x_3-direction) has been given in [11].

The restriction of plane strain, when added to the other kinematical restrictions of incompressibility and fibre-inextensibility, necessarily imposes severe constraints on the possible deformations of the material, and in fact enables us to construct displacement fields with remarkable ease. The associated stress distributions can then be obtained in a relatively simple manner, assuming only very weak restrictions on the stress response of the material.

2.2 Kinematics

The deformation satisfies

$$x_1 = x_1(X_1, X_2)\,, \qquad x_2 = x_2(X_1, X_2)\,, \qquad x_3 = X_3 \tag{14}$$

with

$$\begin{aligned} \underset{\sim}{A} &= \underset{\sim}{A}(X_1, X_2) = \underset{\sim}{i}_1 \cos\Phi + \underset{\sim}{i}_2 \sin\Phi \\ \underset{\sim}{a} &= \underset{\sim}{a}(X_1, X_2) = \underset{\sim}{i}_1 \cos\phi + \underset{\sim}{i}_2 \sin\phi \\ \underset{\sim}{n} &= \underset{\sim}{n}(X_1, X_2) = -\underset{\sim}{i}_1 \sin\phi + \underset{\sim}{i}_2 \cos\phi\,, \end{aligned} \tag{15}$$

where $\underset{\sim}{i}_1$ and $\underset{\sim}{i}_2$ are the base-vectors along the X_1 and X_2 coordinate axes, and Φ and ϕ are the initial and current angles made by the inextensible direction with the X_3-axis ($\phi = 0$). If we treat $\underset{\sim}{a}$ and $\underset{\sim}{n}$ as functions of ϕ, then the derivatives of $\underset{\sim}{a}$ and $\underset{\sim}{n}$ with respect to ϕ are

$$\underset{\sim}{a}'(\phi) = \underset{\sim}{n}(\phi)\,, \qquad \underset{\sim}{n}'(\phi) = -\underset{\sim}{a}(\phi)\,. \tag{16}$$

The curvature of fibres and normal lines are the rates of change of their angles with respect to position:

$$\kappa_a = \frac{1}{r_a} = \frac{\partial\phi}{\partial a} \equiv (\underset{\sim}{a}.\underset{\sim}{\nabla})\phi\,, \qquad \kappa_n = \frac{1}{r_n} = \frac{\partial\phi}{\partial n} \equiv (\underset{\sim}{n}.\underset{\sim}{\nabla})\phi\,. \tag{17}$$

For future reference we note that in plane deformations

$$\begin{aligned} \kappa_a &= a_i \partial\phi/\partial x_i = \cos\phi(\partial\phi/\partial x_1) + \sin\phi(\partial\phi/\partial x_2) \\ &= \frac{\partial}{\partial x_1}(\sin\phi) - \frac{\partial}{\partial x_2}(\cos\phi) = -\frac{\partial n_i}{\partial x_i} \end{aligned} \tag{18}$$

and the Serret-Frenet relations give

$$a_j \frac{\partial a_i}{\partial x_j} = \kappa_a n_i , \qquad a_j \frac{\partial n_i}{\partial x_j} = -\kappa_a a_i ,$$
$$n_j \frac{\partial a_i}{\partial x_j} = \kappa_n n_i , \qquad n_j \frac{\partial n_i}{\partial x_j} = -\kappa_n a_i . \tag{19}$$

The kinematical result of most immediate importance relates to κ_n:

$$\begin{aligned} \kappa_n = n_i \partial\phi/\partial x_i &= -\sin\phi \left(\frac{\partial\phi}{\partial x_1}\right) + \cos\phi \left(\frac{\partial\phi}{\partial x_2}\right) \\ &= \frac{\partial}{\partial x_1}(\cos\phi) + \frac{\partial}{\partial x_2}(\sin\phi) = \frac{\partial a_1}{\partial x_1} + \frac{\partial a_2}{\partial x_2} \\ &= \partial a_i/\partial x_i . \end{aligned} \tag{20}$$

Similarly, the curvature κ_N of the initial normal lines is

$$\kappa_N = \partial A_R/\partial X_R \equiv \underset{\sim}{\nabla}_0 . \underset{\sim}{A} .$$

Thus, from (5) we obtain

$$\kappa_n = \kappa_N ; \tag{21}$$

that is, in plane deformations, the curvature of the orthogonal trajectories of the fibres is constant at any particle.

If the fibres are *initially parallel* (as for closely-packed fibres), though not necessarily straight, then $\underset{\sim}{\nabla}_0 . \underset{\sim}{A}$ is zero, and hence

$$\kappa_n = \kappa_N = 0 , \tag{22}$$

so that *all normal lines are initially straight and remain straight.* This means that in any deformation the fibres form a family of parallel curves (such as parallel, straight lines or concentric circles). So the distance between two fibres, measured along normal lines, is constant along the length of the two fibres. Moreover, for volume preservation (area preservation, in plane strain), *the normal distance between two fibres must remain constant throughout the deformation.*

If the straight $\underset{\sim}{n}$-lines for parallel fibres intersect along some curve in the plane, then $\underset{\sim}{a}$ is discontinuous across this curve. We have already observed that in such a case the curve must bisect the angle between the $\underset{\sim}{a}$-lines on each side of it. If straight $\underset{\sim}{n}$-lines intersect at a single point in the plane, forming a *crease* there, then $\underset{\sim}{a}$ is discontinuous only at that point. This point will then be the focus of a fan region in which the fibres will lie in circular arcs.

Hence for the case of parallel fibres these geometrical conditions are so restrictive and so simple that it is possible, and straightforward, to determine complete states of deformation when the shapes of one or a few fibres are given. It is therefore not surprising that most published solutions of plane problems involve materials with initially parallel fibres, especially since they are the most likely to occur in practice.

Problems in which one of the boundaries lies along a fibre are especially simple. If the deformed shape of the fibre is given, the straight normal lines perpendicular to it can be constructed and the orthogonal trajectories of these straight lines are therefore the loci of the remaining fibres. When the position of one particle on each fibre is given, the positions of the others along the fibre are obtained from the inextensibility condition.

The deformation can therefore be determined completely for the material covered by these normals through the boundary fibre. For those parts of the body not so covered, it is clear that there is no unique kinematically admissible deformation field. Indeed, even for the rest of the body, further information about boundary displacements is required in order to determine the deformation uniquely. This lack of uniqueness is illustrated by the deformations shown in Figure 2; a block, initially rectangular with fibres in the X_2-direction, is deformed so that its side $X_1 = 0$ lies along the polygonal boundary ABC in the figure (the broken

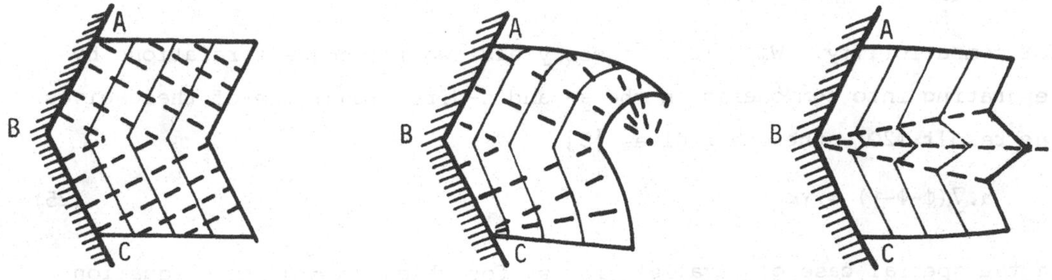

Figure 2. Kinematically admissible deformations of a block satisfying the same boundary displacement constraints

lines are $\underset{\sim}{n}$-lines and the solid curves represent fibres).

The kinematical results discussed above for 'parallel-fibre' materials can be interpreted as simple consequences of the fact that they can deform only by fibres sliding along one another - *simple shear along the* $\underset{\sim}{a}$*-lines* - apart from a rigid body rotation (which is taken into account by the directions of $\underset{\sim}{a}$ and $\underset{\sim}{n}$) and a rigid translation (omitted for convenience). The deformation can be represented by

$$\underset{\sim}{F} = \underset{\sim}{a} \otimes \underset{\sim}{A} + (\underset{\sim}{n}+\gamma \underset{\sim}{a}) \otimes \underset{\sim}{N} + \underset{\sim}{k} \otimes \underset{\sim}{K}, \tag{23}$$

where $\underset{\sim}{k}$ and $\underset{\sim}{K}$ $(= \underset{\sim}{k})$ are the unit vector in the X_3-direction and γ is the *amount of shear*. In fact (23) is also correct [1] for materials with non-parallel fibres, and is easily proved [9] as follows. Consider an infinitesimal cube of edge ds located at $\underset{\sim}{X}$. Then after deformation the edge elements $\underset{\sim}{A}$ ds and $\underset{\sim}{K}$ ds become $\underset{\sim}{a}$ ds and $\underset{\sim}{k}$ ds by inextensibility and plane strain respectively; $\underset{\sim}{N}$ ds maps onto an element of form $(c_1\underset{\sim}{n}+c_2\underset{\sim}{a})$ ds. Incompressibility requires that the deformed volume remain as $(ds)^3$, so $c_1 = 1$; c_2 then identifies the amount of shear γ. Hence we can write

$$d\underset{\sim}{x} = (\underset{\sim}{A}.d\underset{\sim}{X})\underset{\sim}{a} + (\underset{\sim}{N}.d\underset{\sim}{X})(\underset{\sim}{n}+\gamma\underset{\sim}{a}) + (\underset{\sim}{K}.d\underset{\sim}{X})\underset{\sim}{k}$$

or, equivalently,

$$dx_i = a_i A_R dX_R + (n_i+\gamma a_i) N_R dX_R + k_i K_R dX_R$$

which immeditely gives the required result (23).

The deformation is completely specified by γ and the fibre-angle (defining $\underset{\sim}{a}$ and $\underset{\sim}{n}$). They are not independent, but must be such that

$$\partial F_{iR}/\partial X_S = \partial F_{iS}/\partial X_R \tag{24}$$

for compatibility. With (23), these yield two independent relations by separating into components in the $\underset{\sim}{a}$- and $\underset{\sim}{n}$-directions; one of these gives the result (20), the other gives [5]

$$\underset{\sim}{a}.\underset{\sim}{\nabla}(\phi-\Phi-\gamma) = \gamma\kappa_n . \tag{25}$$

In the special case of parallel fibres, for which $\kappa_n = 0$, this equation shows that $\phi-\Phi-\gamma$ is constant along fibres. Thus if the *normal* line through a particle $\underset{\sim}{X}$ makes an angle Φ with some fixed direction (usually the X_2-direction) before deformation and an angle ϕ with the same direction after deformation, then the amount of shear at that particle is

$$\gamma = \phi - \Phi + c\,, \tag{26}$$

where c is a constant on each $\underset{\sim}{a}$-line. This result is particularly useful when considering force resultants later.

2.3 Stress

The stress has the form

$$\sigma_{ij} = Ta_i a_j - p(\delta_{ij} - a_i a_j) + s_{ij}\,. \tag{8}$$

We assume that the material has reflectional symmetry in the planes X_3 = constant; hence

$$\sigma_{13} = s_{13} = 0\,, \qquad \sigma_{23} = s_{23} = 0 \tag{27}$$

in plane deformations. Then, with (9), there are only two independent non-zero components of the extra-stress $\underset{\sim}{s}$. It is clear that one must be the in-plane shearing stress

$$\sigma_{12} = s_{12} = S\,, \qquad \text{say}, \tag{28}$$

and that the other is related to an out-of-plane normal stress difference. A little algebra [2] shows that in fact (8) can be rewritten [1] as

$$\sigma_{ij} = Ta_i a_j - P(\delta_{ij} - a_i a_j) + S(a_i n_j + n_i a_j) + S_3 k_i k_j \tag{29}$$

where T and -P are the *total* normal stresses on surfaces perpendicular to $\underset{\sim}{a}$ and $\underset{\sim}{n}$ respectively, and S_3 $(= \sigma_{33} + P)$ is a normal stress difference. In the more usual matrix notation, the stress can be exhibited as

$$\underset{\sim}{\sigma} = \begin{pmatrix} T & S & 0 \\ S & -P & 0 \\ 0 & 0 & \sigma_{33} \end{pmatrix}\,,$$

where the components are referred to *local* cartesian axes coinciding with the directions $\underset{\sim}{a}$, $\underset{\sim}{n}$ and $\underset{\sim}{k}$ respectively.

In plane strain, the deformation with respect to the local axis is characterised completely by the amount of shear γ. Thus S and S_3 (or equivalently σ_{33}) are determined by the value of γ or its history. For elastic materials, S and S_3 are functions of γ whose forms can be determined from a *single simple shearing experiment*. For viscoelastic or plastic behaviour, if the deformation is plane at all times, S and S_3 are functionals of the history of γ. In all cases we denote this dependence as

$$S = S(\gamma)\ , \qquad S_3 = S_3(\gamma)\ . \tag{30}$$

This constitutive equation - (29) with (30) - is the most general form possible for plane deformations (assuming the reflectional symmetry); its relative simplicity makes it preferable to the much more complicated forms derivable from the three-dimensional forms discussed in Chapter I. Furthermore we shall find that in fact only S is involved in the solution of plane problems - nothing need be said about the form of S_3 except that it is to be independent of X_3.

T and P are determined from the equilibrium equations. The convenient representation is in terms of their components in the $\underset{\sim}{a}$ and $\underset{\sim}{n}$ directions. Following the derivation of (12) from (8), and using the mutual orthogonality of unit vectors $\underset{\sim}{a}$, $\underset{\sim}{n}$ and $\underset{\sim}{k}$, we find that (29) yields

$$\frac{\partial T}{\partial a} + (T+P)\frac{\partial a_i}{\partial x_i} = -n_j\frac{\partial S}{\partial x_j} + S\left(a_i n_j\frac{\partial a_i}{\partial x_j} + \frac{\partial n_i}{\partial x_j} + a_i a_j\frac{\partial n_i}{\partial x_j}\right)\ ;$$

a similar relation follows from (13). Then (17) - (19) finally reduce the equilibrium equations to the form

$$\frac{\partial T}{\partial a} + (T+P)\kappa_n = -\frac{\partial S}{\partial n} + 2\kappa_a S\ , \tag{31}$$

$$\frac{\partial P}{\partial n} - (T+P)\kappa_a = \frac{\partial S}{\partial a} + 2\kappa_n S\ , \tag{32}$$

and $\partial P/\partial x_3 = 0$.

At a boundary point where the unit outward normal is $\underset{\sim}{\nu}$, (29) implies that

$$\sigma_{ij}\nu_j = (T\underset{\sim}{a}.\underset{\sim}{\nu} + S\underset{\sim}{n}.\underset{\sim}{\nu})a_i + (S\underset{\sim}{a}.\underset{\sim}{\nu} - P\underset{\sim}{n}.\underset{\sim}{\nu})n_i\ .$$

This gives boundary conditions on T and P in terms of the applied surface traction $\underset{\sim}{T}^{\nu}$ (say):

$$T = (\underset{\sim}{T}^{\nu}.\underset{\sim}{a}-S\underset{\sim}{n}.\underset{\sim}{\nu})/(\underset{\sim}{a}.\underset{\sim}{\nu}) , \qquad P = (S\underset{\sim}{a}.\underset{\sim}{\nu}-\underset{\sim}{T}^{\nu}.\underset{\sim}{n})/(\underset{\sim}{n}.\underset{\sim}{\nu}) \tag{33}$$

provided the boundary is not a fibre nor a normal line. On boundaries along fibres or normal lines, (33) need not be satisfied; in that case the boundary becomes a singular line carrying a finite force that equilibrates the mis-match between $\underset{\sim}{T}^{\nu}$ and the traction $(\sigma_{ij}\nu_j)$ on the material side of the boundary. If the boundary is curved, this 'surface tension' causes a discontinuity in the normal stress across it. Further discussion is given in a later section.

Equations (31) and (32) immediately show that every kinematically admissible deformation is also statically admissible. For then the fields $\underset{\sim}{a}$ and $\underset{\sim}{n}$ are known, giving κ_a and κ_n directly and S through the relevant constitutive equation: so (31) and (32) are partial differential equations for T and P in characteristic form with the fibres and normal lines as the characteristics. Equation (31) determines the variation of T along a fibre, and (32) governs the variation of P along a normal line. With boundary conditions (33) to give T at one end of a fibre and P at one end of each normal line, the stress field is determinate.

When the material has *parallel* reinforcement, κ_n is zero and equations (32) are further simplified. Unless the deformed fibres are also straight, it is convenient to introduce quasi-polar coordinates (ξ,ϕ) with ξ as the distance between the $\underset{\sim}{a}$-line and a given reference fibre $(\xi = 0)$, and ϕ the fibre-angle labelling the relevant, straight $\underset{\sim}{n}$-line (Figure 3); we assume ξ to increase in the positive $\underset{\sim}{n}$-direction. Since the fibres lie along parallel curves, κ_n is zero and the fibre radius of curvature r must vary linearly along each normal line:

$$r(\xi,\phi) = 1/\kappa_a = r_0(\phi) - \xi , \tag{34}$$

where $r_0(\phi)$ is the radius of curvature of the reference $\underset{\sim}{a}$ line $\xi = 0$. Equations (31) and (32) now take the form

$$\frac{\partial T}{\partial \phi} = 2S - r\frac{\partial S}{\partial \xi} , \qquad P - r\frac{\partial P}{\partial \xi} + T = -\frac{\partial S}{\partial \phi} \tag{35}$$

and can be explicitly integrated to give T and P:

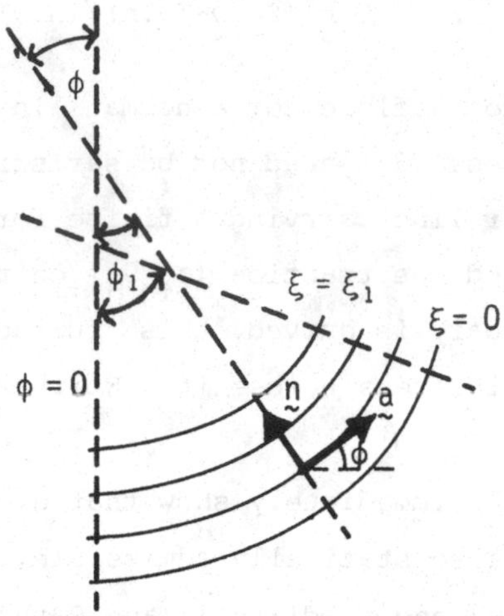

Figure 3. Quasi-polar coordinates ξ, ϕ such that the $\underset{\sim}{a}$-lines and $\underset{\sim}{n}$-lines are coordinate curves

$$T(\xi,\phi) = T_0(\xi) + \int_{\phi_0}^{\phi} (2S - r\partial S/\partial\xi)\, d\phi' \tag{36}$$

$$\{r_0(\phi)-\xi\}P(\xi,\phi) = r_0(\phi)P_0(\phi) + \int_0^{\xi} (T+\partial S/\partial\phi)\, d\xi' . \tag{37}$$

Here $P_0(\phi)$ is the pressure on the reference fibre and $T_0(\xi)$ the tension on the reference normal $\phi = \phi_0$.

If in the deformed body there are regions in which the fibres are not only parallel but also straight, then $r_0(\phi)$ is infinite and both κ_a and κ_n are zero there. The fibres and normal lines can then take the role of cartesian axes with coordinates (s,ξ) and equations (31) and (32), with $a \equiv s$, $n \equiv \xi$, can be integrated directly for T and P:

$$T(s,\xi) = T_0(\xi) - \int_{s_0}^{s} (\partial S/\partial\xi)\, ds' , \tag{38}$$

$$P(s,\xi) = P_0(s) + \int_0^{\xi} (\partial S/\partial s)\, d\xi' . \tag{39}$$

Finally we note that a fan region is a special case of a curved region in which the fibres all lie on concentric arcs, with its apex at the centre with zero radius of curvature. Then (37) implies that usually this apex will be associated with a singular pressure; hence in any problem involving deformation under a point load, it is sensible to consider inserting a fan region with apex at the point of loading. This produces a '*crease*', a point where $\underset{\sim}{a}$ suffers a discontinuous change. Creases can also occur in other circumstances. At a crease the equilibrium equations (35) are not valid since the tension T and shear stress S there are not uniquely defined either in magnitude or direction. In these cases the change in stress across such a point must be determined by considering the equilibrium of the neighbourhood of such a point; the technique is discussed in considerable detail in [12].

2.4 Hyperbolicity and singular stress layers

From the stress solutions (36) and (38) we see that the fibre tension T at any point is related to the tension T_0 at some other specified point on the same fibre, and not to the tension on any other fibre. In other words, each fibre channels information concerning the tension at some point on that fibre to all other points on the fibre. Similarly, for IFRM plane problems, a normal line channels information about the pressure at some point on that normal. The same comments also apply to the case of non-parallel reinforcement.

This stress channelling occurs for elastic as well as non-elastic materials, and is in complete contrast with conventional theories of isotropic elasticity. A similar situation exists with the displacement field; fibre-inextensibility obviously implies that a specified displacement at some point on a fibre carries a strong implication for the displacement of all other points on that fibre. Mathematically, the contrast is a consequence of the hyperbolicity of the governing partial differential equations (31) and (32) as compared with the ellipticity of the conventional equations of elastostatics.

The physical implication of this channelling is that in real strongly anisotropic materials the conditions at a given point of a body strongly affect the conditions found at large distances away, along an a-line or n-line passing through that point. Thus end effects are not now localised to one or two typical diameters' distance from the boundary, but can possibly extend throughout the body. Hence the usual form of St.Venant's principle is now violated in many problems for which it could be assumed in isotropic elasticity.

This feature of stress channelling is analysed and discussed elsewhere (Chapter IV) in the context of the *exact* equations of *linear* anisotropic elasticity, in which the materials are allowed to be both extensible and compressible; the phenomenon is found to be an asymptotically correct prediction for such materials.

The hyperbolic nature of the governing equations also allows stress discontinuities across the $\underset{\sim}{a}$-lines and $\underset{\sim}{n}$-lines (the characteristics) and leads to the remarkable prediction of possible surfaces of infinite stress occurring along and normal to the fibres.

A simple illustration of this phenomenon has already been given in Figure 1. This is a case in which the shear stress is discontinuous across a *straight* fibre in the deformed configuration. In general, if the discontinuity is of amount $S^+ - S^-$ across some fibre $\xi = \xi_0$ then the magnitude $F(s)$ of the associated stress singularity $F(s)\delta(\xi-\xi_0)$ in T is given by (38):

$$F(s) = -\int_{s_0}^{s} (S^+ - S^-)\, ds + F(s_0)\,, \tag{40}$$

where $F(s_0)$ is usually given by the magnitude (often zero as in the case Figure 1) of the singular stress applied at some boundary $s = s_0$.

The more typical case is that of *curved* parallel fibres. If the shear stress S is discontinuous across some fibre $\xi = \xi_0$ then at that fibre

$$\partial S/\partial\xi = (S^+ - S^-)\delta(\xi-\xi_0)\,.$$

Then (36) shows that T involves a contribution of the form

$$T = F_a\delta(\xi-\xi_0)\,, \tag{41}$$

where F_a, the resultant force in the singular fibre, is given by

$$F_a(\phi) = F_{a0} - \int_{\phi_0}^{\phi} (S^+ - S^-)r\, d\phi'\,, \qquad r = r_0(\phi') - \xi_0\,. \tag{42}$$

If the fibre is curved, this 'surface tension' is equilibrated by a pressure discontinuity across it that is found from (37):

$$r(P^+ - P^-) = \int_{\xi_0-0}^{\xi_0+0} T\,d\xi = F_a(\phi)\,.$$

In the same way, shearing stress discontinuities across normal lines produce Dirac deltas in P and discontinuities in T. For example, by using

$$\partial S/\partial\phi = (S^+ - S^-)\,\delta(\phi-\phi_0)$$

in (37) we find that P involves a contribution of the form $F_n\delta(\phi-\phi_0)$ where

$$rF_n(\xi) = r(\phi_0)F_{n0} + \int_0^{\xi} (S^+ - S^-)\,d\xi'\,.$$

Since normal lines are straight in the cases to which (37) applies, the singularity produces no discontinuity in T for 'parallel-fibre' problems.

3. SOLUTIONS OF PLANE BOUNDARY VALUE PROBLEMS

3.1 Force resultants

In problems with sufficient displacement boundary data, it is possible to determine the deformation purely kinematically, without needing to use the equilibrium equations. In mixed problems or purely traction boundary-value problems, the solution for finite deformations is much less straightforward, even for bodies with parallel fibres. However, if the deformation of some $\underset{\sim}{a}$-line is supposed known, then it is often possible with parallel fibres to make a judicious choice (or intelligent guess) of 'straight' and/or 'curved' and/or 'fan' regions to solve the problem, much as one might construct slipline field solutions in plane rigid-plasticity. Hence in such problems the function $r_0(\phi)$, which gives the shape of the reference fibre, becomes the primary function to be determined by the analysis.

In principle, one could carry this function $r_0(\phi)$ as an unknown throughout the analysis, first constructing the deformation field in terms of $r_0(\phi)$, then using (36), (37) and (29) to determine the pointwise stress distribution, and finally inserting the relevant boundary conditions to provide an equation for $r_0(\phi)$. In practice, this is a cumbersome process. A much more convenient method is to use force *resultants* over finite lengths of fibre or normal line.

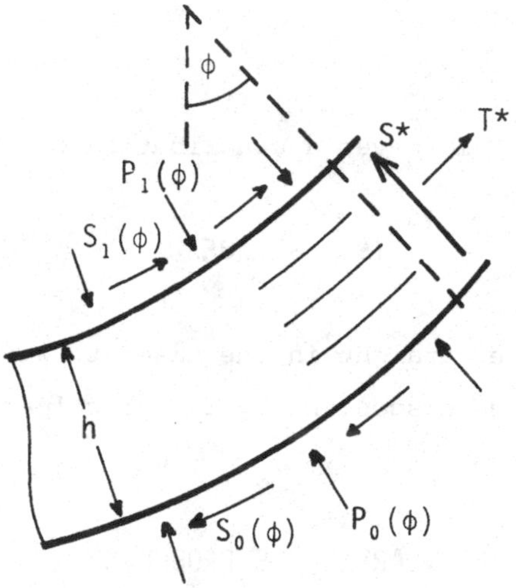

Figure 4. Force resultants

This technique has been found to be particularly effective for problems in which the boundaries of the deformed body are fibre-surfaces. In this case the kinematical result that the normal distance between a pair of parallel fibres is the same all along that pair implies that the body must have a constant thickness. If the 'outer' boundary surface $\xi = 0$ has radius of curvature $r_0(\phi)$, then the inner surface has radius $r_0(\phi)-h$, where h is the thickness (Figure 4). The force resultants $T^*(\phi)$ and $S^*(\phi)$ are defined as the resultant tensile and shearing forces per unit length in the $\underset{\sim}{k}$-direction, acting across a normal line ϕ = constant; hence

$$T^* = \int_0^h T \, d\xi, \qquad S^* = \int_0^h S \, d\xi \, . \tag{43}$$

Integrating the equilibrium equations (35) with respect to ξ between the limits O and h, we obtain

$$\frac{dT^*}{d\phi} = S^* - r_1 S_1 + r_0 S_0 , \qquad T^* = -\frac{dS^*}{d\phi} + r_1 P_1 - r_0 P_0 , \tag{44}$$

where the boundary shear traction and pressure are $S_0(\phi)$, $P_0(\phi)$ on $r = r_0(\phi)$ and $S_1(\phi)$, $P_1(\phi)$ on $r = r_1 = r_0 - h$. Eliminating T^* then results in

$$(P_1 - P_0)\frac{dr_0}{d\phi} + \left(S + \frac{dP_1}{d\phi}\right)(r_0 - h) - \left(S_0 + \frac{dP_0}{d\phi}\right) r_0 = \frac{d^2 S^*}{d\phi^2} + S^* . \tag{45}$$

Now S depends on γ, which in turn depends on $r_0(\phi)$ and ξ, so that in general equation (45) is an integro-differential equation for $r_0(\phi)$. This must usually be solved numerically, with care taken to isolate creases and other singular points at which the radius of curvature may be discontinuous or singular tractions such as concentrated loads are applied.

In some very important cases (45), or equivalently (44), reduces to a simple form which may be solved analytically. Thus, if the preferred, strong direction is *straight* in the undeformed state, and the amount of shear is zero on some normal, equation (26) shows that

$$\gamma = \phi , \tag{46}$$

where that normal is now labelled $\phi_0 = 0$ for convenience. Hence

$$S^* = hS(\phi) ,$$

which reduces (45) to a first order differential equation for $r_0(\phi)$, to be solved in the usual way.

Some simplification of (45) can also be achieved for problems involving fibres which are concentric *circular* arcs in the undeformed configuration. Let R_0 denote the initial radius of the outer boundary fibre, which serves as the reference fibre, and suppose the initial reference normal $\Phi = 0$ remains unsheared at $\phi = 0$. Then inextensibility gives

$$(R_0 - \xi)\Phi = \int_0^{\phi} \{r_0(\phi') - \xi\}\, d\phi' ,$$

so that (26) yields

$$\gamma = I(\phi)/(R_0-\xi) , \tag{47}$$

where $I(\phi)$ is a convenient function defined by

$$I(\phi) = \int_0^{\phi} \{R_0-r_0(\phi')\}\, d\phi' , \qquad r_0(\phi) = R_0 - I'(\phi) . \tag{48}$$

If the response function S is linear in γ - in many cases a very good approximation for materials such as rubber - then for a curved beam of thickness h, we obtain

$$S^* = CI(\phi) , \qquad C = -\mu\, \ell n(1-h/R_0) , \tag{49}$$

where μ is the shear modulus and C is a shear stiffness associated with the beam. Using (49) in (45) then immediately gives a simple, linear second-order differential equation for $I(\phi)$.

3.2 Examples involving initially straight, parallel fibres

Simple shearing

To illustrate the preceding theory in a simple context, we show here the stress analysis supporting the previous assertion that 'the form of $S(\gamma)$ may be determined from a single simple shearing experiment'. The treatment is due to Pipkin [5].

We consider the shearing of an initially rectangular block (Figure 5) of length L and thickness h, which is bonded to rigid plates on its lower and upper surfaces $X_2 = 0$ and $X_2 = h$. The fibres are initially straight and parallel to the X_1-axis. The upper plate is displaced parallel to the X_1-direction with the lower plate held fixed. The end surfaces $X_1 = 0$ and $X_1 = L$ are left free from traction.

In isotropic elasticity theory the solution to this problem is *not* simple shear, and in fact the exact solution has not been found. In IFRM theory, simple shearing is the exact solution.

The kinematical theory is elementary. Knowing that the deformed shapes of the two boundary fibres $X_2 = 0$ and $X_2 = h$ are still straight lines $\xi = 0$, $\xi = h$ parallel to each other, we deduce that the normal lines

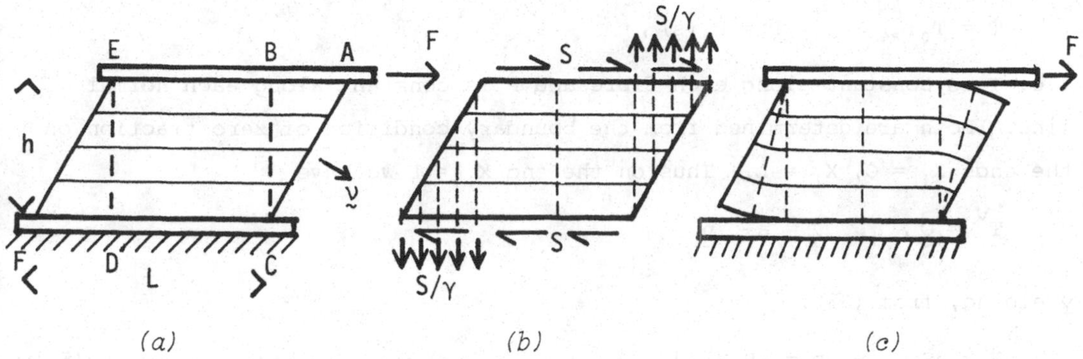

Figure 5. (a) Configuration of the simply sheared block
(b) Tractions supporting simple shearing
(c) Shearing with loss of contact

must be the straight lines s = constant. Hence the remaining fibres, being orthogonal to the normal lines, lie along the straight lines ξ = constant. Since fibre-length and fibre-spacing are conserved, the deformation must therefore have the form

$$s = X_1 + u(X_2)\ , \qquad \xi = X_2\ ; \qquad \underset{\sim}{a} = (1,0,0)\ , \qquad \underset{\sim}{n} = (0,1,0)\ .$$

The amount of shear γ is therefore

$$\gamma = u'(\xi)\ ,$$

so that $S(\gamma)$ is constant along each fibre ξ = constant. The total shearing force along such a line is $LS(\gamma)$, so for equilibrium

$$S(\gamma) = F/L\ ,$$

where F is the total shearing force applied on the top plate. Hence $S(\gamma)$ is a constant, implying that γ is constant so that the deformation is simple shear:

$$s = X_1 + \gamma X_2\ , \qquad \xi = X_2\ ,$$

with the upper plate displaced a distance of γh.

To determine what other forces are required in order to produce this deformation, we note that the relevant stress components are

$$\sigma_{11} = T\ , \qquad \sigma_{12} = S(\gamma)\ , \qquad \sigma_{22} = -P\ .$$

The shear stress σ_{12} is constant, so that equilibrium yields, from (38) and (39),

$$T = T_0(\xi) , \qquad P = P_0(s) ,$$

i.e. T is constant along each fibre and P is constant along each normal line. Both are determined from the boundary condition of zero traction on the ends $X_1 = 0$, $X_1 = L$. Thus on the end $X_1 = L$ we have

$$\underset{\sim}{T}^{\nu} = \underset{\sim}{0} , \qquad \underset{\sim}{\nu} = \underset{\sim}{a} - \gamma \underset{\sim}{n}$$

yielding, from (33),

$$T = \gamma S , \qquad P = -S/\gamma . \tag{50}$$

We note that the fibre-ends being traction-free does *not* imply that T is zero. We note also that the condition at $X_1 = L$ determines $T = \gamma S(\gamma)$ *everywhere*, including the other end $X_1 = 0$. However, the value of P on $X_1 = L$ affects only the triangular portion ABC at the end of the block. The boundary condition at $X_1 = 0$ yields the same values (50); hence there is no contradiction in the solution for T.

In the rectangular region BCDE covered by normal lines that do not cross the two ends, P is indeterminate until its value is specified at one point on each normal line. Hence the *vertical* component of the total force we need to apply to the upper plate is arbitrary, though the distribution of normal traction is not (it must equal S/γ on the section AB of the plate). If the deformation is produced by the horizontal force only, then this component is zero, requiring only that the resultant pressure on BE is $(\gamma h)(S/\gamma) = hS$.

Finally we note that the tensile normal stress $\sigma_{22} = S/\gamma$, acting on AB and DF, may be large enough to cause the bonding between block and plates to fail there. Pipkin [5] has also investigated the case when the bond cannot support a tensile stress. A kinematically admissible solution is shown in Figure 5(c); it can be shown that the boundary conditions can satisfy this deformation, and the stress analysis yields a different F vs. γ relation

$$F = S(\gamma)(L-\gamma h) ,$$

where the upper surface has again been displaced a distance γh.

Finite deflection of a cantilever

One of the first practical applications of the 'ideal' theory was to predict the deformation of a linearly elastic IFRM cantilever under point loading [13]. The analysis is given elsewhere (Chapter III) and is trivially simple. Here we show that the analysis for non-linear behaviour is also simple.

We consider an initially straight cantilever (Figure 6) of length L and thickness h, and reinforced along its length. It is built-in at one end (X_1 = O) and deformed by applying a concentrated load P normal to the upper surface X_2 = h as shown.

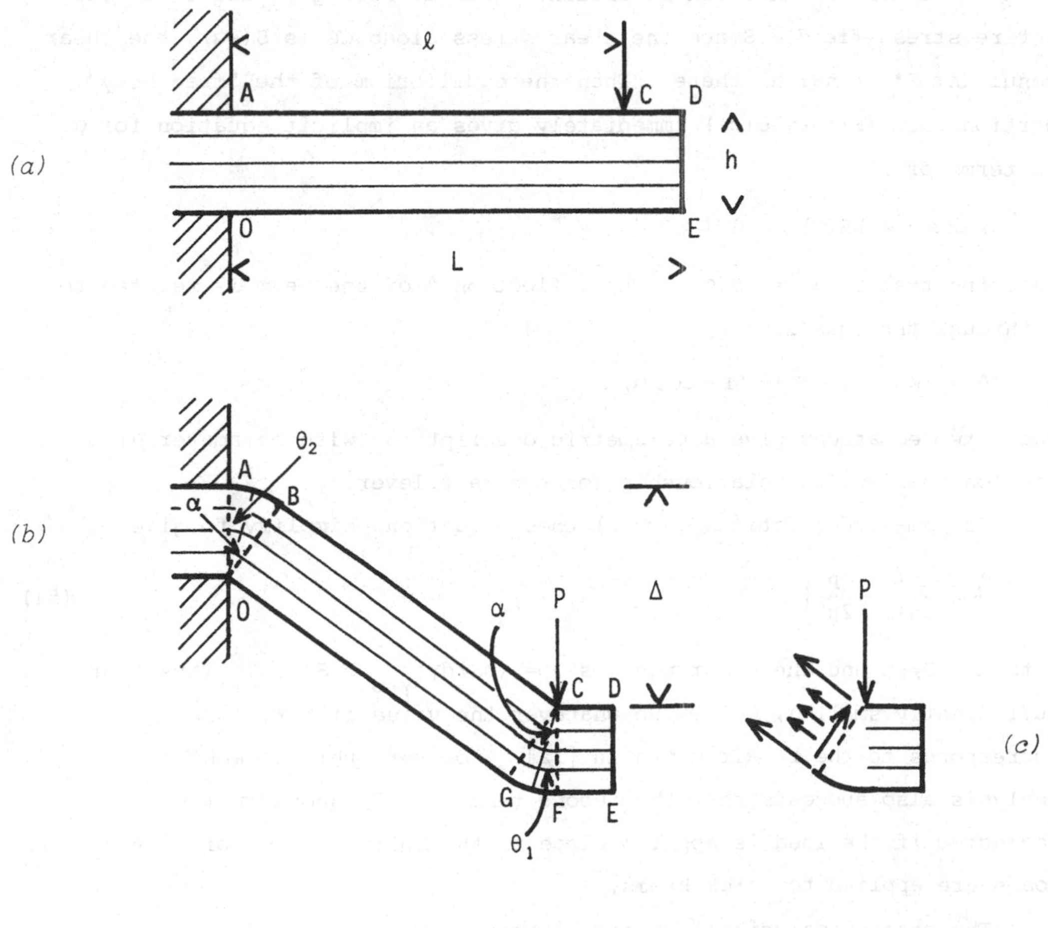

Figure 6. Finite deflection of a cantilever:
(a) undeformed state
(b) deformed configuration
(c) tractions on free body

Since the external tractions are the point load at C and the wall reaction at O, we expect each of these points to be an apex of a fan region. At the same time, the condition of zero traction at the free end requires that DE ($X_1 = L$) must remain a normal line (otherwise there is a non-zero amount of shear, with a consequential load acting in the boundary normal DE). The displacement is zero at the built-in end. A kinematically admissible deformation which satisfies these conditions, and which does not lead to contradiction of any of the boundary conditions, is shown in Figure 6(b).

The deformed configuration is completely determined once the beam angle α is known. This can be obtained without having to solve for the entire stress field. Since the shear stress along CG is $S(-\alpha)$, the shear resultant S^* is $hS(-\alpha)$ there. Then the equilibrium of the 'free body' portion CDEG (Figure 6(c)) immediately gives an implicit equation for α in terms of P:

$$P\cos\alpha = hS(\alpha)\ ,$$

assuming that $S(-\alpha) = -S(\alpha)$. The deflection Δ of the beam is related to α through the equation

$$\Delta = (\ell - h\alpha)\sin\alpha + h(1-\cos\alpha)\ .$$

These two equations give a parametric description (with parameter α) of the load-deflection relationship for the cantilever.

For small deflections ($\alpha \ll 1$) these equations simplify to give

$$\frac{\Delta}{\ell} \sim \frac{P}{\mu h}\left(1 - \frac{P}{2\mu\ell}\right), \tag{51}$$

with $\alpha \sim P/\mu h$ and the shear modulus $\mu = (dS/d\gamma)_{\gamma=0} \equiv S'(0)$. Thus, for sufficiently small α, $\Delta \sim P\ell/\mu h$ whatever the value of h/ℓ; this corresponds to the result given in [12]. However, this elementary analysis also suggests that the second term in (51) should *not* be neglected if the load is applied close to the built-in end, or if end loads are applied to thick beams.

The stress analysis is straightforward. The amount of shear is obviously zero in CDEF and hence on the normal line CF. Thus (26) shows that γ is equal to the fibre angle ϕ everywhere, so throughout the cantilever the amount of shear is

$$\gamma = (0, -\theta_1, -\alpha, -\theta_2) \quad \text{in (CDEF, CFG, GOBC, ABO)}, \tag{52}$$

where each of the general angles θ_1 and θ_2 lies in the range $(0,\alpha)$. Hence the shear stress $S(\gamma)$ is constant along each normal line, so that $\partial S/\partial\xi$ vanishes everywhere inside the cantilever. It is now a simple matter to solve for T and P. We note that the boundary ADEO is comprised entirely of fibres or normal lines, so we must apply the condition of zero surface traction everywhere except C and O with care. For straight portions, the continuity of normal stress, with (33) and $\underset{\sim}{T}^{\nu} = \underset{\sim}{0}$, gives

$$T = 0 \quad \text{on DE},$$

$$P = 0 \quad \text{on BC, CD, EF, GO}.$$

Then, for example, equations (36), (38) and (52) immediately show that

$$T = 2\int_0^{\phi} S(\eta)\, d\eta, \qquad 0 < \xi < h,$$

where ϕ denotes the fibre angle at any point.

Of particular interest are the singular stresses which must occur in the surfaces of the beam by virtue of the shear stress discontinuities there. On the lower boundary fibre ($\xi = 0$) the discontinuity is $S(\phi)$, so that along OG the magnitude of the singular tension is, from (40),

$$F_0(s) = -S(-\alpha)s + F_0(0) = sS(\alpha) + F_0(0). \tag{53}$$

The radius of curvature of the portion of GF is obviously $r = h$ so that, using (53) to give F_0 at G, (42) yields

$$F_0(s) = (\ell-\alpha h)S(\alpha) + F_0(0) - h\int_{-\alpha}^{-\theta_2} S(\eta)\, d\eta$$

along GF, where the fibre-length s measured from O is

$$s = \ell - h\theta_2.$$

The solution for F_0 is completed by evaluating $F_0(0)$ from the boundary condition $F_0 = 0$ at E and hence at F, where $s = \ell$ and $\theta_2 = 0$. So the distribution of singular stress on OE is

$$F_0 = \begin{cases} -(\ell-\alpha h-s)S(\alpha) - h\int_0^{\alpha} S(\eta)\, d\eta & \text{in OG} \\ -h\int_0^{\theta_2} S(\eta)\, d\eta & \text{in GF} \end{cases} .$$

This result shows that the singular stress in the lower surface is entirely compressive, with its maximum occurring at the built-in end. A similar analysis shows that the upper surface suffers tensile singular stress with maximum also at the built-in end.

3.3 Example - squashing of a hollow tube

As a last illustration of the plane IFRM theory, we consider the finite deformation of a body with initially curved, parallel fibres. A pressurised circular tube with azimuthal reinforcement is laterally compressed between two flat plates (Figure 7). Full details of the analysis are given in [13].

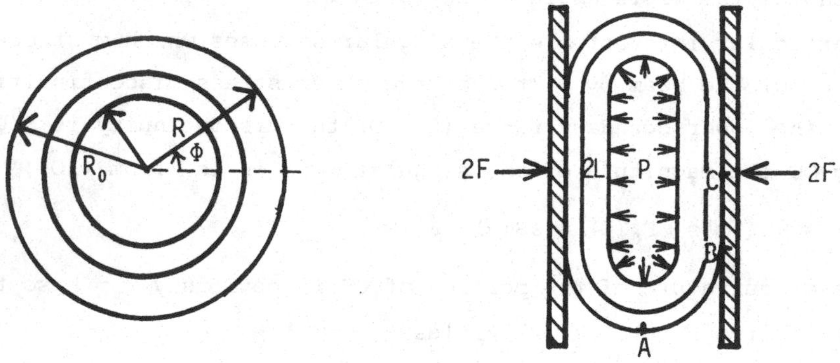

(a) *(b)*

Figure 7. Cross-section of the (a) initial and (b) squashed shapes of a tube

Although this is an extremely complicated mixed boundary-value problem in the context of isotropic elasticity, the IFRM theory reduces the problem to a relatively elementary piece of analysis. The kinematical results of the previous Section 2 suggest the deformed configuration shown in Figure 7(b). It is a simple matter to construct the deformation field once the shape of the outer fibre is determined. This shape is given by solving for the radius of curvature $r_0(\phi)$ of the curved portion of that fibre; the contact length 2L is determined from the inextensibility of AB to be given by

$$L = \int_0^{\frac{1}{2}\pi} r_0(\phi)\, d\phi = \pi R_0 ;$$

equivalently we can write

$$I(\tfrac{1}{2}\pi) = L . \tag{54}$$

Following the analysis in Section 2, equations (45) - (49) show that for $S = \mu\gamma$ the solution is given by

$$r_0(\phi) = R_0 - dI/d\phi ,$$

where $I(\phi)$ satisfies

$$d^2I/d\phi^2 + k^2 I = 0 ,$$

$$k^2 = C/(p+C) , \tag{55}$$

$$C = -\mu\, \ell n(1-h/R_0) .$$

Hence with (54) and $I(0) = 0$ we obtain the solution

$$I(\phi) = L \sin k\phi \operatorname{cosec} \tfrac{1}{2}k\pi \tag{56}$$

so that

$$r_0(\phi) = R_0 - kL \cos k\phi \operatorname{cosec} \tfrac{1}{2}k\pi . \tag{57}$$

The stress analysis is now again straightforward, and the details are omitted. The boundary fibres $\xi = 0$ and $\xi = h$ again carry singular stress due to the discontinuities in shear stress there. It can be shown that the tensile load in the outer boundary fibre decreases monotonically along ABC while that in the corresponding part of the inner boundary fibre increases monotonically. Since none of the fibres actually *cross* a boundary, the fibre stress T and boundary fibre forces are arbitrary to within additive constant stresses $T_e(\xi)$, say, which have a zero force resultant. The stress field is therefore ambiguous to the extent of this tension, constant along each fibre, and the pressure $P_e(\xi,\phi)$ arising as a reaction to it.

There is no ambiguity, however, in the actual deformation corresponding to a given force 2F on the plates. This force can be related to the contact length 2L, and hence to $r_0(\phi)$ and the shape of the tube, by considering the equilibrium of the flat rectangular region adjacent to each plate. The shear resultant across each end is, from (49) and (56),

$$S^*(\tfrac{1}{2}\pi) = CI(\tfrac{1}{2}\pi) = CL ,$$

so equilibrium yields

$$F = (p+C)L . \tag{58}$$

This result shows that $p+C$ can be interpreted as a stiffness coefficient, partly due to internal pressure p and partly due to the shear stiffness C associated with the tube wall. From (55), k^2 is the proportion of stiffness due to the material; the limiting case $k = 0$ corresponds to $C = 0$, the case of a thin membrane, and also to the case of high inflation pressure whether the tube wall is thin or not.

Finally we note that the radius of curvature vanishes at point A ($\phi = 0$) when $r_0(0) = h$, i.e. when the contact length L reaches the critical value L_c where (57) yields

$$L_c = (R_0-h)k^{-1}\sin\tfrac{1}{2}k\pi .$$

This occurs when F reaches the value $(p+C)L_c$. When the load increases further, the analysis gives a negative radius of curvature at A ($\phi = 0$) implying that the boundary fibre is cusped and crosses itself, so giving an unacceptable state of deformation. We therefore introduce a fan region with apex at A, forming a crease there; the analysis is a little more complicated, but still straightforward [13].

3.4 Non-uniqueness and energy considerations

The preceding problem provides an illustration of unusual non-uniqueness present in IFRM solutions of ostensibly perfectly well-posed boundary value problems. It shows that in some cases when fibres (or normals) do not cross the external boundaries the deformation can be completely determined yet the stress field is still ambiguous to the extent of a constant fibre-tension $T_e(\xi)$ and the pressure $P_e(\xi,\phi)$ arising as a reaction to it. This is a potentially serious limitation on the usefulness of the IFRM theory, particularly in its implications for fracture of strongly anisotropic materials. In the example just given, the singular stress layers in the tube walls have obvious physical implications for the performance of loaded radial tyres (with the tube axis interpreted as the direction around the major circumference of the tyre); but each boundary fibre-load is given only to within an arbitrary constant.

This ambiguity could be resolved by formulating a theory of small displacements superposed on large for slightly extensible and compressible materials. The basic large deformation would be that found by using IFRM theory. However, the relevant analysis has yet to be completed.

Another approach could be based on the principles of minimum energy. These have been used very recently to resolve an ambiguity of a different kind, which arise when more than one kinematically and statically amissible deformation exists which satisfy all of the boundary conditions. The alternative deformations differ in whether or not lines of discontinuity are present. Such a case arose in the solution of a machining problem, which was eventually solved [14] using the one solution that had no discontinuity line. The physical reason for choosing the smooth solution has been given by Pipkin [15], who has analysed two IFRM solutions of a bending problem, one continuous and the other having a discontinuity line (Figure 8). Both solutions satisfy all kinematic and equilibrium requirements. The 'continuous' solution shown in Figure 8(a) is the same as the cantilever solution we have already analysed, and Pipkin showed that the total energy of the body and the loading system is a minimum.

In contrast, Pipkin also showed that the discontinuous solution (obtained with similar analysis) does *not* minimise the total energy, nor even render it stationary. In a virtual deformation, some material on one side of the discontinuity line moves through the line to the other side, so that this material suffers a change $\Delta\gamma$ in its amount of shear.

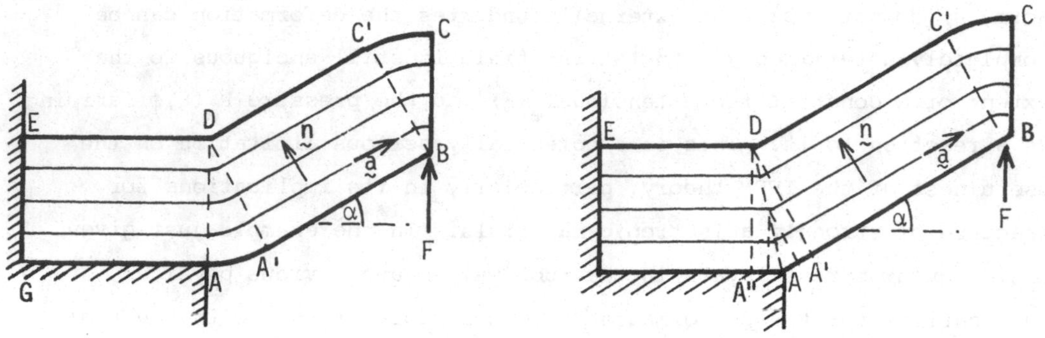

Figure 8. Alternative deformations in a bending problem

The amount of energy created, per unit volume, for this material is

$$e = \Delta W - \sigma_{\nu\tau}\Delta\gamma ,$$

where $\sigma_{\nu\tau}$ is the shear stress and ΔW is the change of stored energy density. ΔW is zero when $\Delta\gamma$ is infinitesimal, but generally not zero when there is a finite discontinuity. Yet Figure 8(b) shows clearly that

$$\Delta\gamma = 2 \tan \tfrac{1}{2}\alpha$$

in this discontinuous case, a *finite* amount even though the displacement is infinitesimal. Thus e need not be zero, so that energy is *not stationary* in small virtual deformations, let alone not being a minimum.

In these problems, uniqueness is restored [9] by requiring solutions to minimise the energy. Discontinuity lines are admissible only when they are imposed by displacement boundary conditions; in such cases the discontinuity line cannot move in any admissible virtual deformation. Hence we should obtain continuous solutions whenever possible.

4. FINITE AXISYMMETRIC DEFORMATIONS

4.1 Introduction

In this sub-section we briefly consider finite axisymmetric

deformations, without twist, of cylindrical bodies uniformly reinforced by fibres initially parallel to the axis (the 'closely-packed' case). Such two-dimensional problems are nearly as simple to analyse as are the plane problems treated in the previous sub-sections. However, complete analytical solutions are less easy to obtain [5] [16], especially for the stress field associated with any kinematically and statically admissible deformation. The theory is readily extended [16] to tubes with initially curved reinforcement.

4.2 Kinematics

We consider deformations in which the particle initially at the place $\underset{\sim}{X}$ with cylindrical coordinates (R,Θ,Z) goes to $\underset{\sim}{x}$ with coordinates (r,θ,z) with

$$r = r(R,Z) , \qquad \theta = \Theta , \qquad z = z(R,Z) .$$

The fibre-directions and normal lines are given by

$$\underset{\sim}{A} = (0,0,1) , \qquad \underset{\sim}{a} = (\sin\phi, 0, \cos\phi) , \qquad \underset{\sim}{n} = (\cos\phi, 0, -\sin\phi) \tag{59}$$

so that

$$\partial\underset{\sim}{a}/\partial\phi = \underset{\sim}{n} , \qquad \partial\underset{\sim}{n}/\partial\phi = -\underset{\sim}{a} .$$

Here ϕ is again the fibre-angle (between $\underset{\sim}{a}$ and the Z-axis).

Since $\underset{\sim}{\nabla}.\underset{\sim}{a}$ is zero for an initially parallel fibre-reinforcement, we obtain

$$0 = \underset{\sim}{\nabla}.\underset{\sim}{a} = \underset{\sim}{n}.\underset{\sim}{\nabla}\phi + (\sin\phi)/r$$

so that, from (17),

$$\kappa_n = -(\sin\phi)/r . \tag{60}$$

Thus the curvature of a normal line is known at every point of the deformed body. Hence if the shape of some one fibre in the deformed body - called the 'boundary fibre', for convenience - is specified, then in principle we can determine the shape of every other fibre that crosses the boundary fibre. A more convenient result is obtained by noting that

$$\underset{\sim}{n}.\underset{\sim}{\nabla}(r\sin\phi) = (\underset{\sim}{n}.\underset{\sim}{\nabla}r)\sin\phi + r\underset{\sim}{n}.\underset{\sim}{\nabla}(\sin\phi)$$

$$= \cos\phi\sin\phi + r\cos\phi\,\underset{\sim}{n}.\underset{\sim}{\nabla}\phi$$

$$= 0$$

from (60). Hence the value of $r\sin\phi$ is constant along any given normal line:

$$r\sin\phi = r^*\sin\phi^* = N, \quad \text{say}, \tag{61}$$

where r^* and ϕ^* are the radial coordinate and fibre-angle of the point at which the normal crosses the boundary line.

If $\phi = 0$ at some point on a normal line, then (61) shows that $\phi = 0$ everywhere on that line, i.e. the normal line is a straight radial line.

Otherwise ϕ is not zero at any point on the line, in which case we note that the points (r,z) on it are related through

$$\frac{dz}{dr} = -\tan\phi = -\frac{N}{r}\left\{1 + \left(\frac{dz}{dr}\right)^2\right\}^{\frac{1}{2}},$$

using (61). By substituting $r = |N|\cosh\psi$, it is easy to obtain the solution

$$z = z^* - N[\cosh^{-1}\{|\operatorname{cosec}\phi|\} - \cosh^{-1}\{|\operatorname{cosec}\phi^*|\}], \tag{62}$$

which is the equation of a catenary. So every normal line is a catenary, the shape of which is determined by the appropriate pair of values z^* and ϕ^* specifying the intersection with the boundary line. When all the normal lines have been constructed, the fibre-lines can be obtained as orthogonal trajectories. This procedure will usually be carried out more easily by graphical methods than by analysis.

Once the field of fibres has been constructed, then the deformation has in fact been completely determined once the position of one point on each fibre is known. Noting that a material curve that is initially a circle of radius R about the symmetry axis becomes a circle of radius $r(R,Z)$, we deduce that an element initially at (R,Z) is stretched in the azimuthal direction in the ratio

$$\lambda = r/R.$$

Superimposed on this stretching is an amount of shear γ between the fibres, just as in plane strain. The deformation gradient tensor $\underset{\sim}{F}$ is therefore similar [16] to that of plane strain, but now involves λ as well as γ:

$$\underset{\sim}{F} = \underset{\sim}{a} \otimes \underset{\sim}{A} + \lambda^{-1} (\underset{\sim}{n} + \gamma \underset{\sim}{a}) \otimes \underset{\sim}{N} + \lambda \underset{\sim}{k} \otimes \underset{\sim}{K} . \tag{63}$$

Now $\underset{\sim}{k}$ (= $\underset{\sim}{K}$) denotes the unit vector in the azimuthal direction. As in plane strain theory, the compatibility relations $\partial F_{iM}/\partial X_N = \partial F_{iN}/\partial X_M$ yield an equation - (60) - for κ_n, and an equation for γ:

$$\frac{\partial}{\partial a} (\gamma - \phi) = \frac{\gamma}{r} \sin \phi . \tag{64}$$

Here $\partial/\partial a$ again denotes differentiation with respect to fibre-length. It is now not possible to integrate (64) directly to give γ in terms of ϕ, as in plane strain. Nevertheless, once the fibre-shapes have been determined, it is possible to obtain γ at every point either by numerically integrating (64) along the fibres or by introducing convenient intermediate functions [16] which reduce the problem to quadratures.

4.3 Stress

The stress must be like that for plane deformations, but with stretching imposed in the asimuthal direction. Thus we have the form

$$\sigma_{ij} = Ta_i a_j - P(\delta_{ij} - n_i n_j) + S(a_i n_j + a_j n_i) + S_3 k_i k_j ,$$

with $\underset{\sim}{k}$ now in the azimuthal direction, and both S and S_3 now depending on λ as well as γ:

$$S = S(\lambda, \gamma) , \qquad S_3 = S_3(\lambda, \gamma) . \tag{65}$$

The equations of equilibrium can be written in forms similar to those of plane strain by again using the appropriate expressions in (12) and (13). However, terms in S_3 now appear [5] in addition to those with S, and κ_n is non-zero even though the reinforcement is initially parallel:

$$\begin{aligned} \frac{\partial T}{\partial a} &= 2\kappa_a S - \frac{\partial S}{\partial n} - \frac{1}{r} (S \cos \phi - S_3 \sin \phi) , \\ \frac{\partial P}{\partial n} &= \kappa_a (T+P) + \frac{\partial S}{\partial a} - \frac{1}{r} (S \sin \phi + S_3 \cos \phi) , \end{aligned} \tag{66}$$

and $\partial P/\partial\theta = 0$.

If a kinematically admissible deformation has been obtained, then S and S_3 can be determined from the appropriate constitutive equations so that all the quantities in these equations except T and P are known. Just as in plane strain, the first equation determines T to within an arbitrary constant along each fibre, and the second then gives P to within an arbitrary constant along each normal line. So every kinematically admissible axisymmetric deformation of such an IFRM body is also statically admissible.

Equations (66) are two hyperbolic partial differential equations with the fibres and normal lines as the characteristics. All the relevant remarks made in the plane strain theory also hold for these axisymmetric deformations. In particular, singular stress layers can, and often do, occur in fibre surfaces of revolution and in normal (annular) surfaces. The theory is very similar to that of plane strain.

Although the analysis described here appears to be relatively simple, the two equations (66) will ordinarily have to be integrated numerically. Only two analytical solutions have been presented [5] for mixed boundary-value problems - axial shearing and pure radial expansion of a tube. Neither of these involves singular stress layers.

REFERENCES

[1] PIPKIN, A.C. and ROGERS, T.G., *J. Appl. Mech.* 38 (1971) 634-640

[2] SPENCER, A.J.M., *Deformations of Fibre-reinforced Materials*, Clarendon Press, Oxford, 1972

[3] GREEN, A.E. and ADKINS, J.E., *Large Elastic Deformations*, Clarendon Press, Oxford, 1960

[4] MULHERN, J.F., ROGERS, T.G. and SPENCER, A.J.M., *Proc. Roy. Soc. Lond.* A301 (1967) 473-492

[5] PIPKIN, A.C., Finite deformations of ideal fibre-reinforced composites, in *Composite Materials, Vol.2: Micromechanics*,

Sendeckyj, G.P., Ed., Academic Press, New York (1973) 251-308

[6] ROGERS, T.G., Finite deformations of strongly anisotropic materials, in *Theoretical Rheology*, Hutton, J.F., Pearson, J.R.A. and Walters, K., Eds., Applied Science Pub., London (1975) 141-168

[7] ROGERS, T.G., *Rheol. Acta* 16 (1977) 123-133

[8] PIPKIN, A.C., Finite deformations in materials reinforced with inextensible cords, in *Finite Elasticity*, Rivlin, R.S., Ed., ASME AMD 27 (1977) 91-102

[9] PIPKIN, A.C., *Advances in Applied Mechanics* 19 (1979) 1-51

[10] SPENCER, A.J.M., Stress concentration layers in finite deformation of fibre-reinforced elastic materials, *Proc. IUTAM Symposium on Finite Elasticity, Bethlehem 1980*, Carlson, D.E. and Shield, R.T., Eds. Martinus Nijhoff, The Hague (1982) 357-377

[11] PIPKIN, A.C., *Q. Appl. Math.* 32 (1974) 253-263

[12] ROGERS, T.G. and PIPKIN, A.C., *J. Appl. Mech.* 38 (1971) 1047-1048

[13] ROGERS, T.G. and PIPKIN, A.C., *Q. J. Mech. Appl. Math.* 24 (1971) 311-330

[14] EVERSTINE, G.C. and ROGERS, T.G., *J. Composite Mat.* 5 (1971) 94-106

[15] PIPKIN, A.C., *Q. Appl. Math.* 35 (1978) 455-463

[16] PIPKIN, A.C., *Q. J. Mech. Appl. Math.* 28 (1975) 271-284

III
PLANE PROBLEMS FOR FIBRE-REINFORCED LINEARLY ELASTIC SOLIDS

A. H. ENGLAND

Department of Theoretical Mechanics
University of Nottingham
Nottingham, NG7 2RD
England

1. INTRODUCTION

In this section we examine plane problems for fibre-reinforced materials. The constraints which were described in the previous sections may be built into linear elasticity to yield very simple systems of equations. We first consider an incompressible material which is reinforced by a single family of inextensible fibres which are parallel to the x_1-axis. The constraints are

INCOMPRESSIBILITY $$\frac{\partial u_1}{\partial x_1} + \frac{\partial u_2}{\partial x_2} + \frac{\partial u_3}{\partial x_3} = 0\,, \tag{1}$$

INEXTENSIBILITY $$\frac{\partial u_1}{\partial x_1} = 0\,. \tag{2}$$

If the material is undergoing a *plane strain* deformation parallel to the (x_1,x_2)-plane so that

$$\left.\begin{array}{l} u_1,\ u_2 \text{ are independent of } x_3 \quad \text{and} \\ u_3 \equiv 0\,, \end{array}\right\} \tag{3}$$

then, from (1) and (2),

$$\frac{\partial u_2}{\partial x_2} = 0 \tag{4}$$

and hence

$$u_1 = u_1(x_2)\ , \qquad u_2 = u_2(x_1)\ . \tag{5}$$

For such systems it is more explicit to employ the usual Cartesian coordinate notation

$$(u_1,u_2,u_3) = (u,v,w)\ ,$$

$$(x_1,x_2,x_3) = (x,y,z)\ ,$$

so that the combination of incompressibility, inextensibility parallel to the x-direction and plane strain as described above yields the simple form of the displacement field:

$$u = u(y)\ , \qquad v = v(x)\ . \tag{6}$$

The constitutive relations are given by equation (28) of Chapter I. The only explicit relations are

$$\sigma_{xy} = \mu\{u'(y)+v'(x)\}\ , \tag{7}$$

where $\mu = \mu_L$, and $\sigma_{xz} = \sigma_{yz} = 0$ by the plane-strain assumptions. The remaining stress components are determined in terms of the reactions to the constraints of incompressibility and inextensibility and are

$$\sigma_{xx} = -p+T\ , \qquad \sigma_{yy} = \sigma_{zz} = -p\ . \tag{8}$$

These must satisfy the equations of equilibrium, which reduce to

$$\begin{aligned} &\frac{\partial\sigma_{xx}}{\partial x} + \mu u''(y) = 0\ , \\ &\frac{\partial\sigma_{yy}}{\partial y} + \mu v''(x) = 0\ , \\ &\frac{\partial\sigma_{zz}}{\partial z} = 0\ , \end{aligned} \tag{9}$$

using (7). Hence, on integration,

$$\sigma_{xx} = -\mu x u''(y) + F(y) ,$$
$$\sigma_{yy} = -\mu y v''(x) + G(x) , \tag{10}$$

where F(y) and G(x) are arbitrary functions of the variables shown. In addition,

$$\sigma_{zz} = \sigma_{yy} . \tag{11}$$

The very restrictive system of equations (6), (7) and (10) determines the displacement and stress fields for plane-strain deformations of an incompressible material reinforced by a family of straight inextensible fibres.

An (almost) identical system of equations arises in two other contexts. Consider an incompressible material reinforced by a single family of inextensible fibres which are parallel to the y-axis. Again, for plane-strain deformations parallel to the (x,y)-plane, the kinematic constraints reduce to (6). The only non-zero shear stress is (7) and the normal stress components satisfy

$$\sigma_{yy} = -p + T , \qquad \sigma_{xx} = \sigma_{zz} = -p . \tag{12}$$

When these are substituted into the equilibrium equations, the relations (10) are found together with relation $\sigma_{zz} = \sigma_{xx}$. Hence, apart from the out-of-plane stress component, these equations are identical with those obtained above.

Similarly, consider a compressible material reinforced by two orthogonal families of identical inextensible fibres which are aligned parallel to the x- and y-axes. For plane-strain deformations in the (x,y)-plane then

$$\frac{\partial u}{\partial x} = 0 , \qquad \frac{\partial v}{\partial y} = 0 ,$$

and u and v are independent of z, so that

$$u = u(y) , \qquad v = v(x) ,$$

which is equation (6) again. Note that since $w \equiv 0$, $\operatorname{div} \underset{\sim}{u} = 0$ so that no volume changes take place in this compressible material. The constitutive equation must contain reactions to these inextensibility constraints and has the components (from equations (39) and (40) of Chapter I)

$$\sigma_{xy} = \mu_L\{u'(y)+v'(x)\}, \qquad \text{where} \quad \mu_L = \mu + \mu_1 + \mu_2 ,$$
$$\sigma_{zz} = 0 . \tag{13}$$

The normal stress components σ_{xx} and σ_{yy} depend on the reaction stresses along the fibre directions and are determined from the equilibrium equations to have the forms (10).

In the next section, we examine the solution of plane-strain boundary-value problems for these materials. We shall describe them in terms of an incompressible material reinforced by a single family of fibres aligned along the x-direction. Pipkin and Rogers [1] refer to this material as an *IDEAL* fibre-reinforced material. The two alternative interpretations discussed above should also be borne in mind.

2. PLANE STRAIN PROBLEMS FOR AN IDEAL FIBRE-REINFORCED MATERIAL

The basic equations are (6), (7) and (10), namely

$$u = u(y) , \qquad v = v(x) ,$$
$$\sigma_{xy} = \mu\{u'(y)+v'(x)\} ,$$
$$\sigma_{xx} = -\mu x u''(y) + F(y) , \tag{14}$$
$$\sigma_{yy} = -\mu y v''(x) + G(x) .$$

These equations have a much simpler character than those of anisotropic elasticity and consequently many boundary-value problems of difficulty in anisotropic elasticity may be solved with comparative ease using this limiting case of an ideal fibre-reinforced material.

2.1 Displacement boundary-value problems

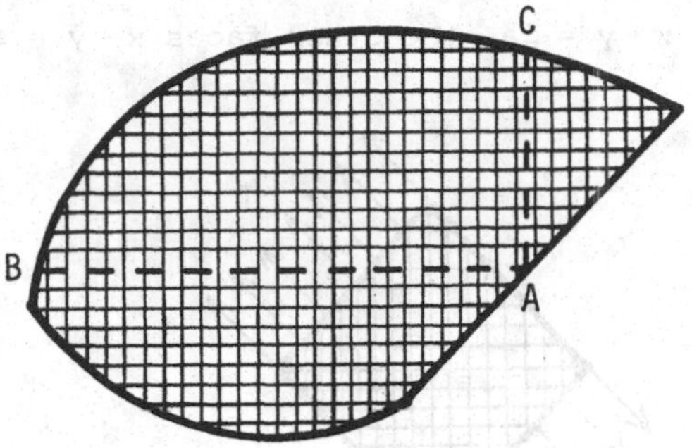

Figure 1

Suppose the displacement is specified at each point on the boundary of the body shown in *Figure 1*. Then since the displacement field satisfies the very restrictive equations $(14)_1$ the u-component of displacement at A must be identical to that at B, and the v-component of displacement at A must be identical to that at C. Only if the displacement boundary conditions satisfy these compatibility conditions for each point A of the boundary will it be possible to find the displacement field in the body. In this case the displacement field $u = u(y)$, $v = v(x)$ is defined explicitly and hence from $(14)_2$ the shear stress is known. It is not possible to define σ_{xx} and σ_{yy} since they are undetermined to within a constant stress F(y) along each fibre or G(x) along each normal line respectively. The boundary conditions do not yield enough information to define these quantities.

2.2 A stress boundary-value problem

The general solution to the stress boundary-value problem for a body which is symmetrical about an axis parallel to the fibre direction has been given by England [2]. First we consider a simple example for which the solution is known to the corresponding problem for an anisotropic elastic body.

Square reinforced along its diagonals

Consider the square shown in Figure 2 under the action of a uniform tension over the faces $x+y = \pm a$, the other faces $x-y = \pm a$ being unstressed.

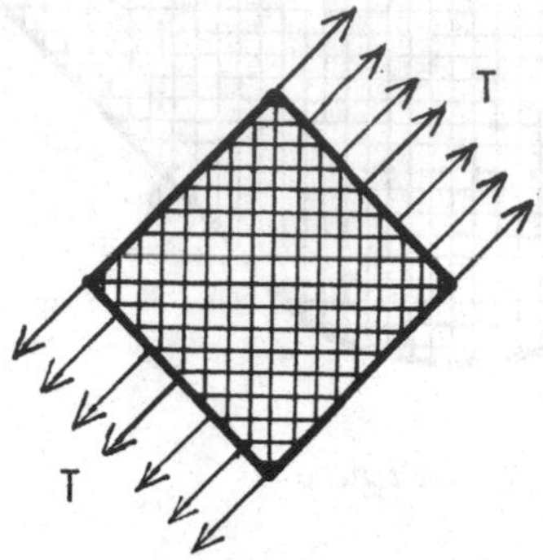

Figure 2

Let us consider the equilibrium of the portion of the body to the left of line $x = x_0$. Then, resolving in the y-direction, we find

$$\int_{-(a+x_0)}^{a+x_0} \sigma_{xy}\, dy - T(x_0+a) = 0, \qquad x_0 < 0. \tag{15}$$

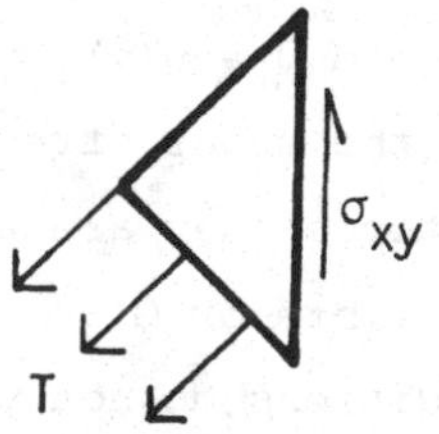

Figure 3

On substituting for σ_{xy} from $(14)_2$ and integrating we find

$$\mu[u(a+x_0)-u\{-(a+x_0)\}] + 2\mu(a+x_0)v'(x_0) = T(a+x_0), \qquad -a < x_0 \leqslant 0. \tag{16}$$

It is easy to confirm that in the range $0 \leqslant x_0 \leqslant a$ the corresponding equation is

$$\mu[u(a-x_0)-u\{-(a-x_0)\}] + 2\mu(a-x_0)v'(x_0) = T(a-x_0). \tag{17}$$

Similarly if we examine the equilibrium of the portion of the body below $y = y_0$, then the resultant force in the x-direction gives

$$\int_{-(a+y_0)}^{(a+y_0)} \sigma_{xy}\, dx - T(a+y_0) = 0, \qquad y_0 < 0, \tag{18}$$

see Figure 4, and hence

$$2\mu(a\pm y)u'(y) + \mu[v(a\pm y)-v\{-(a\pm y)\}] = T(a\pm y), \qquad y \lessgtr 0 \tag{19}$$

(dropping the suffix on y).

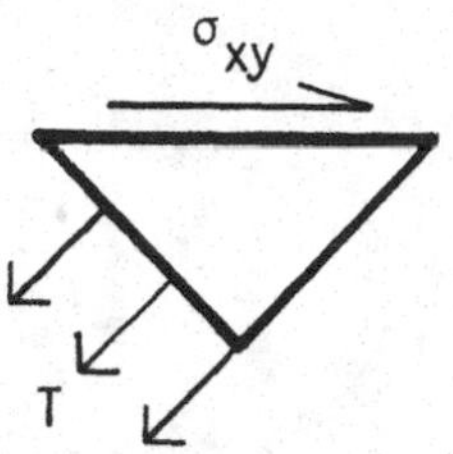

Figure 4

The solution of the differential-difference equations (17) and (19) is simply

$$u = \frac{T}{4\mu}y, \qquad v = \frac{T}{4\mu}x$$

and this is unique to within a rigid body displacement. The stress field is such that $\sigma_{xy} = \frac{1}{2}T$ from $(14)_2$ and since $\sigma_{xx} = F(y)$, $\sigma_{yy} = G(x)$, the boundary conditions yield

$$\sigma_{xx} = \sigma_{yy} = \tfrac{1}{2}T. \tag{21}$$

This method of solution may be generalised to more involved shapes and boundary conditions.

Let us consider an orthotropic elastic material with axes of symmetry along the coordinate directions and satisfying the plane strain boundary-value problem shown in Figure 2. Then the stress field is precisely that given above, namely

$$\sigma_{xx} = \sigma_{yy} = \sigma_{xy} = \tfrac{1}{2}T. \tag{22}$$

The constitutive relations for the orthotropic material are

$$\begin{aligned}\sigma_{xx} &= \ell_{11}e_{xx} + \ell_{12}e_{yy}\,,\\ \sigma_{yy} &= \ell_{12}e_{xx} + \ell_{22}e_{yy}\,,\\ \sigma_{xy} &= 2\mu e_{xy}\,,\end{aligned} \tag{23}$$

where elastic constants ℓ_{ij} satisfy $\ell_{11}\ell_{22} > \ell_{12}^2$ for a positive-definite strain energy function. Thus the displacement field in this anisotropic body satisfies the equations

$$\ell_{11}\frac{\partial u}{\partial x} + \ell_{12}\frac{\partial v}{\partial y} = \tfrac{1}{2}T\,,$$

$$\ell_{12}\frac{\partial u}{\partial x} + \ell_{22}\frac{\partial v}{\partial y} = \tfrac{1}{2}T\,,$$

$$\mu\left(\frac{\partial u}{\partial y} + \frac{\partial v}{\partial x}\right) = \tfrac{1}{2}T\,,$$

which have the solution

$$\begin{aligned} u &= \tfrac{1}{2}T\left(\frac{\ell_{22}-\ell_{12}}{\ell_{11}\ell_{22}-\ell_{12}^2}\right)x + \frac{T}{4\mu}y\,,\\ v &= \frac{T}{4\mu}x + \tfrac{1}{2}T\left(\frac{\ell_{11}-\ell_{12}}{\ell_{11}\ell_{22}-\ell_{12}^2}\right)y\,.\end{aligned} \tag{24}$$

Now, in terms of Young's modulus E' and Poisson's ratio ν' for plane strain extensions of a material having identical properties about the x- and y-axes,

$$\ell_{11} = E'/(1-\nu'^2) = \ell_{22}\,, \qquad \ell_{12} = \nu'\ell_{11}$$

and hence the deformation field (24) has the form

$$u = \frac{T}{4\mu}y + \frac{T(1-\nu')}{2E'}x\,, \qquad v = \frac{T}{4\mu}x + \frac{T(1-\nu')}{2E'}y\,. \tag{25}$$

Hence, when the Young's modulus E' is large compared with the shear modulus μ, the displacement field (25) is well approximated by the solution (20) for the ideal material. This limiting behaviour will be examined in greater detail in subsequent sections.

2.3 The general stress boundary-value problem

The method of solution employed in the last section may be generalised to solve this problem for a body which is symmetrical about an axis parallel to the fibre direction and for which the fibres and normal lines only intersect the boundary twice, see Figure 5.

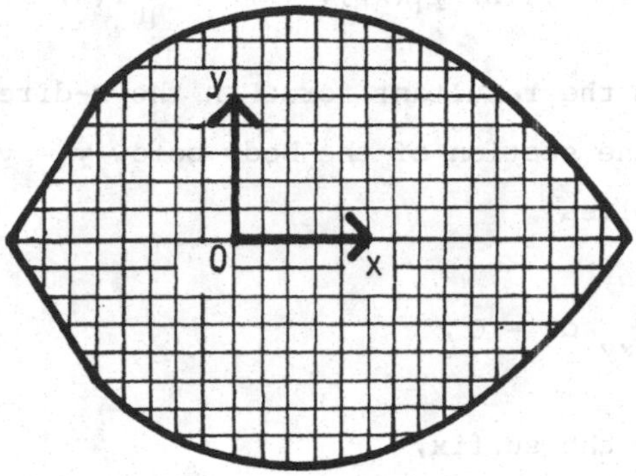

Figure 5

Let us take the axis of symmetry to be y = 0 and position the origin at some convenient point on this axis. Then the boundary of the convex region has the representations

$$y = y_2(x)\,,\quad y > 0\,;\qquad x = x_2(y)\,,\quad x > 0\,;$$
$$y = y_1(x)\,,\quad y < 0\,;\qquad x = x_1(y)\,,\quad x < 0\,; \tag{26}$$

and we note that from symmetry

$$y_1(x) = -y_2(x)\,,\qquad \text{all } x\,, \tag{27}$$

$$x_2(-y) = x_2(y)\,,\qquad x_1(-y) = x_1(y)\,. \tag{28}$$

Perhaps the easiest way to derive the equations satisfied by the displacements is to examine the equilibrium of sections of the body. If we denote the resultant force in the y-direction of the applied tractions on the section of the body to the left of the line $x = x_0$ by $Y(x_0)$, then equilibrium in the y-direction implies

$$Y(x_0) + \int_{y_1(x_0)}^{y_2(x_0)} \sigma_{xy}\, dy = 0 .$$

Hence

$$Y(x_0) + \mu[u\{y_2(x_0)\} - u\{y_1(x_0)\} + \{y_2(x_0) - y_1(x_0)\}v'(x_0)] = 0 ,$$

or, on dropping the suffix,

$$u\{y_2(x)\} - u\{y_1(x)\} + \{y_2(x) - y_1(x)\}v'(x) = -\frac{1}{\mu}Y(x) . \tag{29}$$

Similarly, if $X(y_0)$ is the resultant force in the x-direction of the applied tractions on the section of the body below $y = y_0$, the equilibrium in the x-direction implies

$$X(y_0) + \int_{x_1(y_0)}^{x_2(y_0)} \sigma_{xy}\, dx = 0 ,$$

and hence, on dropping the suffix,

$$\{x_2(y) - x_1(y)\}u'(y) + v\{x_2(y)\} - v\{x_1(y)\} = -\frac{1}{\mu}X(y) . \tag{30}$$

It is necessary to solve these simultaneous ordinary differential equations for $u(y)$ and $v(x)$ throughout the region. In view of the symmetry conditions (27) and (28), let us eliminate $v'(x)$ from these equations by differentiating (30) and substituting from (29). To make this substitution we require the relations.

$$\begin{aligned} y_2\{x_2(y)\} &= |y| , \qquad & y_2\{x_1(y)\} &= |y| , \\ y_1\{x_2(y)\} &= -|y| , \qquad & y_1\{x_1(y)\} &= -|y| , \end{aligned} \tag{31}$$

and hence we find $u(y)$ satisfies the difference-differential equation

$$\begin{aligned} &\{x_2(y) - x_1(y)\}u''(y) + \{x_2'(y) - x_1'(y)\}u'(y) \\ &\qquad - \{x_2'(y) - x_1'(y)\}\left\{\frac{u(|y|) - u(-|y|)}{|y|}\right\} \\ &\qquad\qquad = -\frac{1}{\mu}X'(y) + \frac{1}{2\mu|y|}[Y\{x_2(y)\}x_2'(y) - Y\{x_1(y)\}x_1'(y)] . \end{aligned} \tag{32}$$

Note $x_2(y) - x_1(y)$ is the width of the region at height y and it is convenient to put

$$x_2(y) - x_1(y) = t(y) . \tag{33}$$

In view of the form of (32), let us express u(y) as the sum of an odd and an even function

$$u(y) = u_0(y) + u_e(y)$$

where

$$u_0(-y) = -u_0(y) , \qquad u_e(-y) = u_e(y) . \tag{35}$$

Then (32) becomes

$$t(y)\{u_0''(y)+u_e''(y)\} + t'(y)\{u_0'(y)+u_e'(y)\} - \frac{t'(y)}{|y|}u_0(|y|)$$

$$= -\frac{1}{\mu}X'(y) + \frac{1}{2\mu|y|}[Y\{x_2(y)\}x_2'(y) - Y\{x_1(y)\}x_1'(y)] . \tag{36}$$

On replacing y by -y and noting from the symmetry assumptions (27) and (28) that t(y) = t(-y), equation (36) separates into the equations

$$t(y)u_e''(y) + t'(y)u_e'(y) = -\frac{1}{2\mu}\{X'(y)+X'(-y)\} , \tag{37}$$

and

$$t(y)u_0''(y) + t'(y)u_0'(y) - \frac{t'(y)}{|y|}u_0(|y|)$$

$$= -\frac{1}{2\mu}\{X'(y)-X'(-y)\} + \frac{1}{2\mu|y|}[Y\{x_2(y)\}x_2'(y) - Y\{x_1(y)\}x_1'(y)]$$

$$= R(y) , \tag{38}$$

which may be regarded as holding for $y \geqslant 0$. It will be seen that equation (37) may be integrated immediately to give

$$t(y)u_e'(y) = -\frac{1}{2\mu}\{X(y)-X(-y)\} + \text{constant} \tag{39}$$

and hence $u_e(y)$ may be found. In fact, by replacing y by -y in (39) it can be shown that the constant of integration is zero. The integration of (38) may be easily performed once it is noted that the homogeneous equation

$$t(y)u_0''(y) + t'(y)u_0'(y) - \frac{t'(y)}{y}u_0(y) = 0$$

has the solution $u_0(y) = y$. We can now look for a solution of (38) of the form

$$u_0(y) = yU(y) , \tag{40}$$

from which we find $U(y)$ satisfies the equation

$$yt(y)U''(y) + \{2t(y)+yt'(y)\}U'(y) = R(y) .$$

However, this equation has the integrating factor y and hence

$$\frac{d}{dy}\{y^2t(y)U'(y)\} = yR(y)$$

which integrates to give

$$U'(y) = \frac{1}{y^2t(y)}\int_0^y \eta R(\eta)\, d\eta + \frac{C}{y^2t(y)} .$$

Thus we find

$$u_0(y) = y\int \frac{1}{y^2t(y)}\int_0^y \eta R(\eta)\, d\eta\, dy + Cy\int \frac{1}{y^2t(y)}\, dy + Dy , \tag{41}$$

for $y > 0$, where C and D are constants and $R(y)$ is given in terms of the stress resultants by (38).

The constant C should be chosen so that $u_0(y) = 0$ which ensures that $u(y)$ is continuous at $y = 0$. Now having determined $u(y) = u_0(y) + u_e(y)$ (to within an arbitrary rigid body motion) equation (29) may be integrated to determine $v(x)$. The arbitrary constants occurring in these integrations correspond to a rigid body motion and can be chosen so that the displacement and rotation at the origin are zero by putting

$$u(0) = 0 , \qquad v(0) = 0 ,$$

$$u'(0) - v'(0) = 0 .$$

In this way the displacement field may be completely determined. However, the stress field is not immediately determined on differentiating the displacement field as it contains the arbitrary tensions $F(y)$ and $G(x)$ in (14). These may be found on matching with the boundary conditions and this is illustrated in the following examples.

Circular cross-section

Suppose the cross-section is a circle of radius a then the boundary has the equations

$$y_2 = (a^2-x^2)^{\frac{1}{2}}, \qquad x_2 = (a^2-y^2)^{\frac{1}{2}},$$

$$y_1 = -(a^2-x^2)^{\frac{1}{2}}, \qquad x_1 = -(a^2-y^2)^{\frac{1}{2}},$$

and the thickness $t(y) = x_2(y) - x_1(y) = 2(a^2-y^2)^{\frac{1}{2}}$. In this case the simultaneous equations for the displacements (29) and (30) take the form

$$u\{(a^2-x^2)^{\frac{1}{2}}\} - u\{-(a^2-x^2)^{\frac{1}{2}}\} + 2(a^2-x^2)^{\frac{1}{2}}v'(x) = -\frac{1}{\mu}Y(x),$$

$$2(a^2-y^2)^{\frac{1}{2}}u'(y) + v\{(a^2-y^2)^{\frac{1}{2}}\} - v\{-(a^2-y^2)^{\frac{1}{2}}\} = -\frac{1}{\mu}X(y). \qquad (42)$$

From the solutions (39) and (41) derived in the previous section, $u(y) = u_0(y) + u_e(y)$ where

$$2(a^2-y^2)^{\frac{1}{2}}u_e'(y) = -\frac{1}{2\mu}\{X(y)-X(-y)\},$$

so that

$$u_e(y) = -\frac{1}{4\mu}\int \frac{X(y)-X(-y)}{(a^2-y^2)^{\frac{1}{2}}}\,dy.$$

Similarly

$$u_0(y) = y\int \frac{1}{2y^2(a^2-y^2)^{\frac{1}{2}}}\int_0^y \eta R(\eta)\,d\eta\,dy + Cy\int \frac{1}{2y^2(a^2-y^2)^{\frac{1}{2}}}\,dy + Dy, \qquad (43)$$

where

$$R(y) = -\frac{1}{2\mu}\{X'(y)-X'(-y)\} - \frac{1}{2\mu|y|}[Y\{(a^2-y^2)^{\frac{1}{2}}\} + Y\{-(a^2-y^2)^{\frac{1}{2}}\}]\frac{y}{(a^2-y^2)^{\frac{1}{2}}}. \qquad (44)$$

The resultant forces $Y(x_0)$ and $X(y_0)$ are respectively the y-component of the applied tractions over the sector $-a \leqslant x \leqslant x_0$ and the x-component over the sector $-a \leqslant y \leqslant y_0$. Using polar coordinates $x_0 = a\sin\theta'$, $y_0 = a\cos\theta'$ these have the form

$$Y(x_0) = \int_{\theta'}^{2\pi-\theta'} \{\sigma_{rr}(\theta)\sin\theta + \sigma_{r\theta}(\theta)\cos\theta\} a\, d\theta\,,$$

$$X(y_0) = \int_{-\pi-\theta'}^{\theta'} \{\sigma_{rr}(\theta)\cos\theta - \sigma_{r\theta}(\theta)\sin\theta\} a\, d\theta\,, \tag{45}$$

in terms of the specified normal and shear stress $\sigma_{rr}(\theta)$ and $\sigma_{r\theta}(\theta)$ on the boundary.

For any symmetric system of stress boundary values for which

$$\sigma_{rr}(\theta) = \sigma_{rr}(-\theta) = \sigma_{rr}(\theta+\pi)\,,$$

$$\sigma_{r\theta}(\theta) = -\sigma_{r\theta}(-\theta) = \sigma_{r\theta}(\theta+\pi)\,, \tag{46}$$

it is apparent that

$$Y(x_0) = 0\,, \qquad X(y_0) = 0\,,$$

and hence

$$u_e(y) = 0\,,$$

$$u_0(y) = Cy\int \frac{1}{2y^2(a^2-y^2)^{\frac{1}{2}}}\,dy + Dy\,, \qquad y > 0$$

$$= -\frac{C}{2a^2}(a^2-y^2)^{\frac{1}{2}} + Dy\,, \qquad y > 0\,.$$

For continuity of $u_0(y)$ we require $u_0(0) = 0$ and hence $C = 0$. Thus on integrating (29) we find

$$v = -Dx \qquad \text{and} \qquad u = Dy\,.$$

These displacements merely correspond to a rigid body rotation so without loss of generality we can put $D = 0$. Hence this ideal fibre-reinforced body will support a symmetric stress system without deformation.

Obviously there is a stress field in the disc which is such that

$$\sigma_{xy} = \mu\{u'(y)+v'(x)\} = 0\,.$$

From (14) the forms of the stresses are

$$\sigma_{xx} = F(y)\,, \qquad \sigma_{yy} = G(x)\,,$$

and on substituting into the boundary conditions we find

$$F(y) = \sigma_{rr}(\theta) - \sigma_{r\theta}(\theta)\tan\theta\,,$$

$$G(x) = \sigma_{rr}(\theta) + \sigma_{r\theta}(\theta)\cot\theta\,,$$

thus completing the solution to the problem. It will be noted that the symmetries (46) assumed for the stress components ensure that the boundary values at each end of a fibre (or perpendicular line) define the same value for the stress components; for example, at $\pi - \theta$, $y = a\sin(\pi-\theta) = a\sin\theta$,

$$F(y) = \sigma_{rr}(\pi-\theta) - \sigma_{r\theta}(\pi-\theta)\tan(\pi-\theta)$$

$$= \sigma_{rr}(\theta) - \sigma_{r\theta}\tan\theta\,.$$

A similar result holds for $G(x)$.

It follows directly from this analysis that if the symmetric system of point forces

$$a\sigma_{rr}(\theta) = T\{\delta(\theta-\alpha)+\delta(\theta+\alpha-\pi)+\delta(\theta-\alpha+\pi)+\delta(\theta+\alpha)\}$$

is applied to the disc it will be supported by the stress system

$$\sigma_{xx} = \{\delta(y - a\sin\alpha) + \delta(y + a\sin\alpha)\}T\cos\alpha\,,$$
$$\sigma_{yy} = \{\delta(x - a\cos\alpha) + \delta(x + a\cos\alpha)\}T\sin\alpha\,, \qquad (47)$$

in the disc in which the fibres $y = \pm a\sin\alpha$ and the lines $x = \pm a\cos\alpha$ carry finite forces and the disc remains undeformed. Under this type of loading the material behaves rather like a frame, see Figure 6.

As the only mode of deformation is a shear any unsymmetric system of loading will give rise to a non-zero displacement field. Let us consider the effect of equal and opposite point forces on the composite. Take

$$a\sigma_{rr}(\theta) = T\delta(\theta-\alpha) + T\delta(\theta-\pi-\alpha)\,, \qquad (48)$$

then

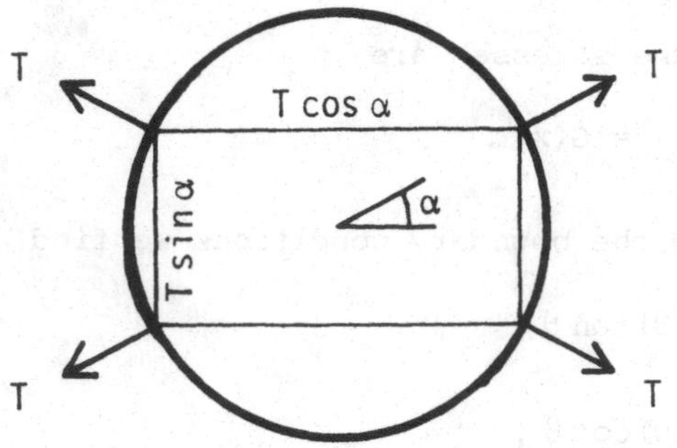

Figure 6

$$Y(x) = -T\sin\alpha\{H(x + a\cos\alpha) - H(x - a\cos\alpha)\} \tag{49}$$

and

$$X(y) = -T\cos\alpha\{H(y + a\sin\alpha) - H(y - a\sin\alpha)\} \tag{50}$$

and these are both even functions. Consequently on examining the differential equations (42) we find

$$u\{(a^2-x^2)^{\frac{1}{2}}\} - u\{-(a^2-x^2)^{\frac{1}{2}}\} + 2(a^2-x^2)^{\frac{1}{2}}v'(x) = \begin{cases} 0, & |x| > a\cos\alpha, \\ -\dfrac{T\sin\alpha}{\mu}, & |x| < a\cos\alpha, \end{cases}$$

$$2(a^2-y^2)^{\frac{1}{2}}u'(y) + v\{(a^2-y^2)^{\frac{1}{2}}\} - v\{-(a^2-y^2)^{\frac{1}{2}}\} = \begin{cases} 0, & |y| > a\sin\alpha, \\ -\dfrac{T\cos\alpha}{\mu}, & |y| < a\sin\alpha. \end{cases}$$

The solution to these equations may be derived by substituting in the general solution (39) and (41), but it may be easily confirmed that in this case u(y) and v(x) are *odd* functions and

$$u(y) = \begin{cases} \dfrac{T}{4a\mu}y, & 0 < y < a\sin\alpha, \\ \dfrac{T\sin\alpha}{2\mu} - \dfrac{T}{4a\mu}y, & a\sin\alpha < y < a, \end{cases}$$

$$v(x) = \begin{cases} \dfrac{T}{4a\mu}x, & 0 < x < a\cos\alpha, \\ \dfrac{T\cos\alpha}{2\mu} - \dfrac{T}{4a\mu}x, & a\cos\alpha < x < a. \end{cases} \tag{51}$$

This result depends on the particular shape of the boundary. This field is illustrated in Figure 7.

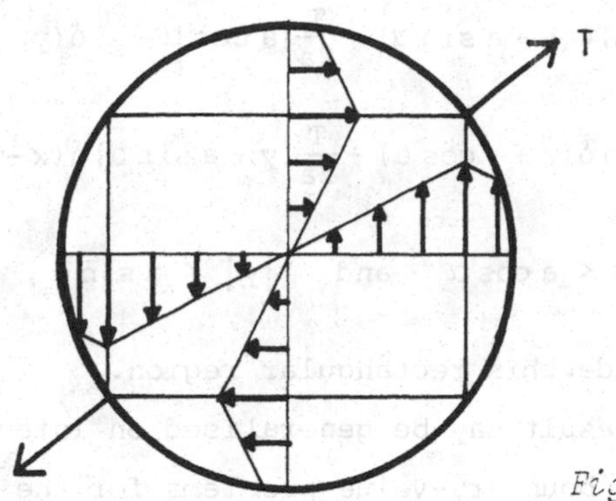

Figure 7

The stress field may now be determined. As $\sigma_{xy} = \mu\{u'(y)+v'(x)\}$ we find

$$\sigma_{xy} = \frac{T}{2a} \quad \text{when} \quad |y| < a\sin\alpha \quad \text{and} \quad |x| < a\cos\alpha$$

and is zero outside this region. Using the relations (14), namely

$$\sigma_{xx} = -\mu x u''(y) + F(y)\,, \qquad \sigma_{yy} = -\mu y v''(x) + G(x)\,,$$

and matching with the boundary conditions, we can find the form of the arbitrary tensions F(y) and G(x). As the boundary conditions have the form

$$\begin{aligned} \sigma_{xx} &= \sigma_{rr}(\theta) - \sigma_{r\theta}(\theta)\tan\theta - \sigma_{xy}\tan\theta\,, \\ \sigma_{yy} &= \sigma_{rr}(\theta) + \sigma_{r\theta}(\theta)\cot\theta - \sigma_{xy}\cot\theta\,, \end{aligned} \tag{52}$$

it is apparent that $F(y) = 0$ when $|y| \neq a\sin\alpha$ and $G(x) = 0$ when $|x| \neq a\cos\alpha$. Clearly there must be infinite stresses (finite forces) carried along the lines $|y| = a\sin\alpha$, $|x| = a\cos\alpha$. Either by a substitution into (52) or (more directly) by an examination of the equilibrium of the fibre along $y = a\sin\alpha$ the finite force on the fibre can be shown to be

$$\sigma_{xx} = \frac{T}{2a}(x + a\cos\alpha)\,\delta(y - a\sin\alpha)\,.$$

Hence the stress field in the disc has the form

$$\sigma_{xx} = \frac{T}{2a}(x + a\cos\alpha)\,\delta(y - a\sin\alpha) + \frac{T}{2a}(a\cos\alpha - x)\,\delta(y + a\sin\alpha)\ ,$$

$$\sigma_{yy} = \frac{T}{2a}(a\sin\alpha - y)\,\delta(x + a\cos\alpha) + \frac{T}{2a}(y + a\sin\alpha)\,\delta(x - a\cos\alpha)\ , \qquad (53)$$

$$\sigma_{xy} = \frac{T}{2a}\,, \qquad |x| < a\cos\alpha \quad \text{and} \quad |y| < a\sin\alpha\ ,$$

where σ_{xy} is zero outside this rectangular region.

This point force result may be generalised on integration to give the results to more general boundary-value problems for the circular disc.

2.4 Three-point bending of a beam

This problem was analysed by Pipkin and Rogers [1] and highlights many of the effects present in strongly anisotropic elastic materials which are rather unexpected when viewed in the context of isotropic elasticity.

Consider the problem shown in Figure 8.

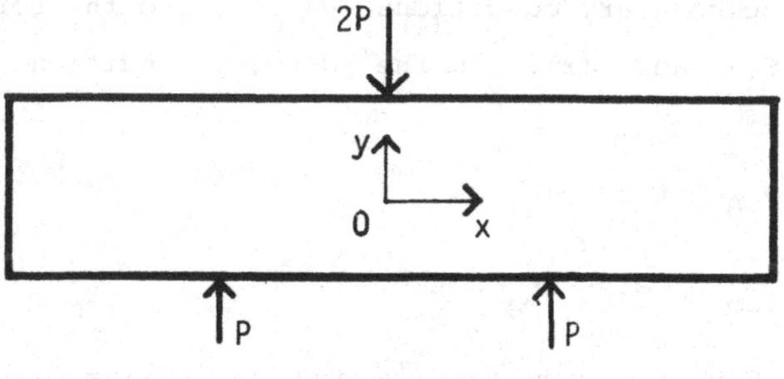

Figure 8

Let the fibres lie along the beam. Let the beam be $|y| \leqslant h$, $|x| \leqslant \ell$ and let the point forces by applied at $x = \pm a$ on $y = -h$.

The displacement field is

$$u = u(y)\ , \qquad v = v(x)\ ,$$

and since it is symmetrical about $x = 0$ we must conclude $u(y) = 0$. If we look at equilibrium in the y-direction for the portion of the beam to the left of the section with abscissa x we find

$$\int_{-h}^{h} \sigma_{xy}\, dy = -Y(x) ,$$

where Y(x) is the resultant applied force in this direction. But $\sigma_{xy} = \mu v'(x)$ since $u(y) = 0$, and hence

$$2\mu h v'(x) = \begin{cases} 0 , & -\ell \leqslant x \leqslant -a , \\ -P , & -a \leqslant x \leqslant 0 , \\ P , & 0 \leqslant x \leqslant a , \\ 0 , & a \leqslant x \leqslant \ell . \end{cases}$$

Hence the beam adopts the sheared configuration shown in Figure 9.

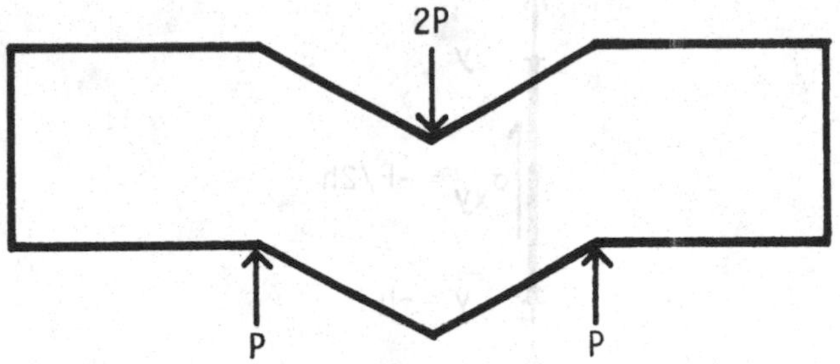

Figure 9

The shear stress is zero in the ranges $a \leqslant |x| \leqslant \ell$ and has the values

$$\sigma_{xy} = \mp \frac{P}{2h} \quad \text{in} \quad \begin{cases} -a \leqslant x \leqslant 0 , \\ 0 \leqslant x \leqslant a , \end{cases} \tag{54}$$

respectively. Thus there are discontinuities in the shear stress across the boundary fibres and normal lines $x = 0$, $x = \pm a$. These may be equilibrated by allowing finite forces to act along each line.

For example, on $y = h$ in $-a \leqslant x \leqslant 0$, equilibrium of a section of the

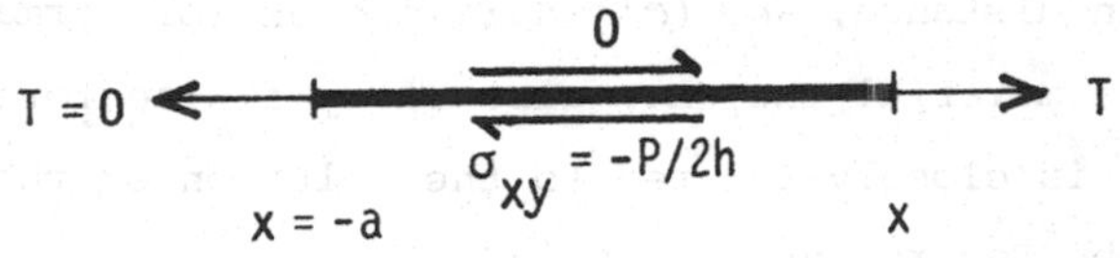

Figure 10

boundary fibre implies (see Figure 10)

$$T = -p(a+x)/2h\,, \qquad -a \leqslant x \leqslant 0\,. \tag{55}$$

Thus the force on this fibre is compressive and increases linearly from zero at $x = -a$ to $-pa/2h$ at $x = 0$. Similarly, on the lower edge $y = -h$, the force in the fibre is tensile and has the value

$$T = p(a+x)/2h\,, \qquad -a \leqslant x \leqslant 0\,. \tag{56}$$

The normal line $x = -a$ also carries a finite force reaction to the inextensibility constraint. From Figure 11 we see

$$T = -p + p(y+h)/2h\,. \tag{57}$$

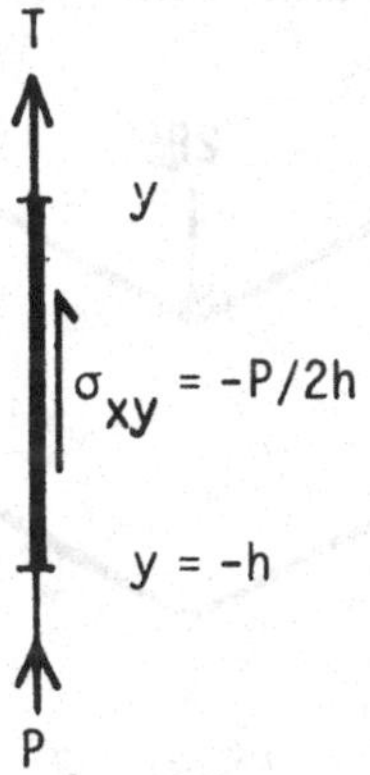

Figure 11

This is a compressive force which decays from $-p$ at $y = -h$ to zero at $y = h$. Similarly it may be shown that the central normal line $x = 0$ carries the compressive load

$$T = -2p + p(h-y)/h \tag{58}$$

which decreases to zero at the lower surface. A symmetrical system of forces acts in $x > 0$. Note *(a)* that the ends of the beam are unstressed, *(b)* where fibres emanate from point forces they carry finite forces which decay linearly with distance, and *(c)* where fibres (or normal lines) form the boundary of the material they may also carry finite forces.

This solution is closely related to the solution of the cantilever problem discussed by Rogers and Pipkin [3].

3. HOLES AND CRACKS IN IDEAL FIBRE-REINFORCED MATERIALS

The problems discussed thus far in this chapter have been restricted to bodies for which the fibres and normal lines intersect the boundary twice only (and in addition the body has an axis of symmetry along or normal to a fibre). This excludes multiply-connected plane regions but the methods we have employed are easily extended to cover most cases of interest.

Rectangular region containing a line crack

Consider a line crack of length 2a in a rectangular plate. Take the origin at the centre of the crack and let the plate have the dimensions $|x| \leqslant \ell$, $|y| \leqslant h$ so that the crack is symmetrically located in the plate (we shall see later that this condition can be relaxed). We shall assume that the crack and the edges of the plate are parallel or normal to the fibres, see Figure 12.

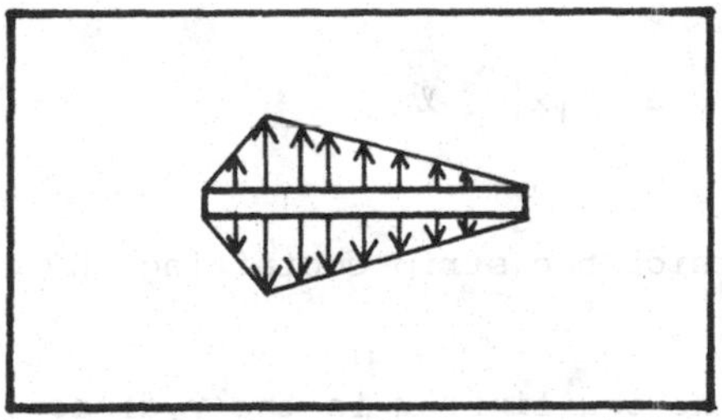

Figure 12

Let us suppose the crack is opened by equal and opposite normal pressure distributions $\sigma_{yy} = -p(x)$, $|x| \leqslant a$ on the upper and lower surfaces of the crack with zero shear stress components. Without loss of generality we can suppose the external edges of the region are unstressed since boundary-value problems for the complete rectangle are easily solved using the methods of Section 2.

Since the problem is symmetrical about y = 0 we must assume $v(x,y) = -v(x,-y)$, but v depends only on x and hence we can conclude

$$v(x) = 0, \qquad a \leqslant |x| \leqslant \ell, \tag{59}$$

with equal and opposite normal displacements over the crack. Note in addition that since $v(x)$ must be continuous

$$v(a) = v(-a) = 0. \tag{60}$$

The problem may now be solved by examining the equilibrium of portions of the body. Consider the portion below the ordinate y, then equilibrium in the x-direction implies

$$\int_{-\ell}^{\ell} \sigma_{xy} \, dx = 0$$

which yields

$$2\ell u'(y) + \int_{-a}^{a} v'(x) \, dx = 0 \tag{61}$$

which implies $u'(y) = 0$, all y, from (59) and (60). Hence $u(y) \equiv 0$ and the the sections $a \leqslant |x| \leqslant \ell$ (the regions outside the lines bounding the crack) remain undeformed when the crack inflates. In fact, since the edges of the region are unstressed, from (14) we can conclude

$$\left.\begin{aligned} \sigma_{xx} &= G(x) = 0, \\ \sigma_{yy} &= F(y) = 0, \end{aligned}\right\} \qquad a < |x| \leqslant \ell, \tag{62}$$

and hence the regions outside the strip containing the crack are completely unstressed.

Now let us examine the equilibrium in the y-direction of the portion of the body in $y > 0$ to the left of abscissa x where $-a \leqslant x < a$ (see Figure 13).

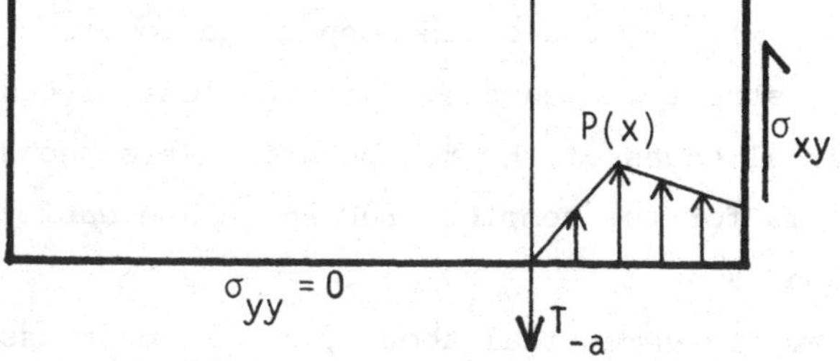

Figure 13

We have shown $\sigma_{yy} = 0$ in $-\ell \leqslant x < -a$; let us suppose the edge line $x = -a$ carries a finite tension T_{-a} at $y = 0$ (and similarly that the edge line $x = a$ will carry the finite tension T_a at $y = 0$). Then equilibrium in the y-direction requires

$$-T_{-a} + \int_{-a}^{x} p(x)\, dx + \int_{0}^{h} \mu v'(x)\, dy = 0$$

so that

$$\mu h v'(x) = T_{-a} - \int_{-a}^{x} p(x)\, dx , \qquad -a < x < a . \tag{63}$$

Now equilibrium of the portion of the region $y > 0$ to the left of x when x lies in the range $a < x < \ell$ implies

$$-T_{-a} - T_a + \int_{-a}^{a} p(x)\, dx = 0 \tag{64}$$

and this can only be achieved if we admit the presence of these singular fibres at the tips of the line crack.

Integration of (63) gives

$$\mu h v(x) = T_{-a}(x+a) - \int_{-a}^{x}\int_{-a}^{x} p(t)\, dt\, dx , \tag{65}$$

since $v(-a) = 0$, and since $v(a) = 0$ the value of T_{-a} may be established:

$$2aT_{-a} = \int_{-a}^{a}\int_{-a}^{x} p(t)\, dt\, dx = \int_{-a}^{a} (a-t)p(t)\, dt , \tag{66}$$

and T_a can be found from (64).

Thus the displacement field is known throughout the region $y > 0$ and hence, by symmetry, everywhere. The shear stress in $y > 0$, $|x| \leqslant a$ is

$$\sigma_{xy} = \frac{1}{h} T_{-a} - \frac{1}{h} \int_{-a}^{x} p(x)\, dx \tag{67}$$

and is independent of y. The normal stress components are (from (14))

$$\sigma_{xx} = F(y) , \qquad \sigma_{yy} = \frac{y}{h} p(x) + G(x)$$

and the boundary conditions on $y = 0$, $|x| < a$ and along the lines $|x| = a$ imply

$$\sigma_{xx} = 0 , \qquad \sigma_{yy} = \left(\frac{y}{h} - 1\right) p(x) , \qquad y > 0 , \qquad |x| < 0 . \tag{68}$$

Once again we must note that σ_{xy} is discontinuous across the edges of the region $0 < y < h$, $|x| < a$ (which is bounded by fibres or normal lines) and hence these four lines must carry finite tensions or compressions.

Consider the bounding line $x = -a$ for $y > 0$. At $y = 0$ there exists a tension T_{-a}, on the side $x = -a-$ the shear stress σ_{xy} is zero and on the side $x = -a+$ the shear stress has the value (67)

$$\sigma_{xy} = \frac{1}{h} T_{-a} .$$

Hence equilibrium of the length y of this line implies that the tension at height y is

$$T\Big|_{x=-a} = T_{-a} - \frac{y}{h} T_{-a} , \qquad 0 \leqslant y \leqslant h , \tag{69}$$

so that the tension decreases linearly from the crack tip to the edge of the plate. Similarly we can show that on the bounding line $x = a$ the tension is

$$T\Big|_{x=a} = T_a \left(1 - \frac{y}{h}\right) \qquad \text{where} \qquad T_a = \frac{1}{2a} \int_{-a}^{a} (a+x) p(x) \, dx . \tag{70}$$

Along the upper face of the crack ($y = 0+$) $\sigma_{xy} = \mu v'(x)$, but the imposed boundary condition is $\sigma_{xy} = 0$, so that the fibre line $y = 0$ must carry a finite force. This force must be zero at the crack tips and hence from equilibrium considerations

$$T\Big|_{y=0+} = - \int_{-a}^{x} \sigma_{xy} \, dx = -\mu v(x) , \qquad |x| \leqslant a . \tag{71}$$

Similarly on the outer edge $y = h$, $|x| \leqslant a$ the boundary condition is $\sigma_{xy} = 0$ but on $y = h-$ the solution yields $\sigma_{xy} = \mu v'(x)$, so that the finite tension along this edge fibre is

$$T\Big|_{y=h} = \mu v(x) . \tag{72}$$

The solution given here has been derived without any assumptions regarding the form of $p(x)$. The stress field for the region $y > 0$ corresponding to inflation of the crack by a constant pressure $p(x) = p_0$ is summarised in Figure 14. Note that

$$T_a = T_{-a} = ap_0 ,$$

$$v(x) = \frac{p_0}{2\mu h}(a^2-x^2) , \qquad |x| \leqslant a , \tag{73}$$

so that the crack inflates into a parabolic shape. The upper and lower faces of the crack are under compressive forces whilst the outer edges are under the tensile force

$$T\Big|_{y=h} = \frac{p_0}{2h}(a^2-x^2) ; \tag{74}$$

the stresses and forces not shown on Figure 14 are zero.

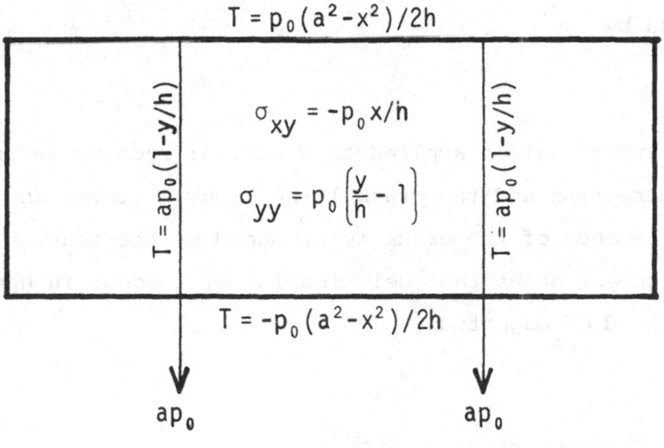

Figure 14

The influence of external forces on this crack configuration is equally simple to evaluate. In particular if a uniform tension $\sigma_{yy} = T_0$ is applied to the surfaces $y = \pm h$ the stress field is simply the above solution (with $p_0 = T_0$) added to the uniform stress field $\sigma_{yy} = T_0$, $\sigma_{xx} = \sigma_{xy} = 0$.

The nature of this solution is quite different from that found in isotropic elasticity (or even orthotropic elasticity) where the crack inflates into an elliptical shape and there are square-root type singular stresses at the crack tips. In addition, for an ideal material the effect of the crack is limited to the region bounded by the crack tip normal lines which is certainly not the case in linear elasticity. However, we might expect this solution to predict the most highly stressed regions of the composite and to yield quantitative estimates of the strength of these stress concentrations. The quantities should be useful in predicting both the onset and mode of failure in a highly anisotropic component.

Suppose a uniform state of stress $\sigma_{yy} = T$, $\sigma_{xx} = \sigma_{xy} = 0$ is applied to the edges of a rectangular plate with fibres along the y-direction, and containing a central crack of width 2a (or edge cracks of width a). The analysis shows that the edge fibres will carry a corresponding load of aT. In the experiment the applied stress is gradually increased until fracture along the cracks is observed when $T = T_F$. Then a reasonable (if tentative) estimate of the failure strength along the fibres in any situation is given by

$$F_F = aT_F \, .$$

If the same experiment is applied to a similar rectangular plate with fibres in the x-direction and the crack lying along a fibre, and delamination at the ends of the crack is produced by the tension $\sigma_{yy} = T_N$, then a similar argument shows that delamination will occur in general if a finite tensile load of magnitude

$$F_N = aT_N$$

is generated normal to the fibres.

Finally if in a simple shear test the composite fails in shear when the shear stress reaches $\sigma_{xy} = S$, then we may expect a similar failure to occur at any point where the shear stress achieves the value S.

By comparing these tensile and shear failure strengths with the stresses present in a given configuration, we may predict the mode of failure. For example, if we consider a rectangular plate under a uniform tension T along the fibre direction with a central crack of width 2a normal to the fibres, then there will be a tensile fibre fracture at the crack tips if $aT > F_F$. Similarly there will be a tensile delamination fracture at the centre points $x = 0$, $y = \pm h$ on the external surface if $a^2T/2h > F_N$. There will be shear failure at all points along the edge lines $x = \pm a$ if $aT/h > S$. We note that an increase in length h of the plate will decrease the possibility of the latter two failures, but that failure is independent of the width of the plate. Hence there will be failure by crack extension if

$$F_F < 2(h/a)F_N, \qquad F_F < Sh . \tag{75}$$

There will be surface cracking at the centre points if

$$F_N < \tfrac{1}{2}(a/h)F_F, \qquad F_N < \tfrac{1}{2}aS . \tag{76}$$

There will be a shear failure along the fibres $x = \pm a$ if

$$S < h^{-1}F_F, \qquad S < 2a^{-1}F_N . \tag{77}$$

Furthermore, we note that when $F_N \simeq F_F$, there will be surface cracking only if $h < \frac{1}{2}a$, that is when the length of the plate is small compared with the crack length. Normally $h \gg a$, in which case the mode of failure will be fracture at the crack tip if $F_F < Sh$. Otherwise if $F_F > Sh$ there will be shear failure. Note that failure by crack extension will lead to the complete rupture of the body whereas the other failure modes merely cause material degradation.

Solutions for internal and external cracks in symmetrical bodies composed of ideal fibre-reinforced materials have been given by England and Rogers [4] and for holes and cracks in rectangular plates and beams by England and Thomas [5] and Thomas [6]. In particular England and Thomas give the solution to the angled crack in a rectangular plate; this solution is non-trivial algebraically but may be obtained by a direct application of the methods described in this section

The solution of such crack and hole problems for finite regions in linear elasticity is beyond the scope of current analysis and requires numerical computation. A solution of this type has been given by Bowie and Freese [7]. Solutions are available, however, for infinite regions of anisotropic material (see Bishop [8]) and these, combined with a boundary-layer analysis (Everstine and Pipkin [9], Spencer [10]), allow comparison to be made between such solutions. These aspects will be discussed in Chapters IV and VI.

4. PLANE STRAIN AND GENERALISED PLANE STRESS PROBLEMS FOR FIBRE-REINFORCED MATERIALS

Previous sections in this chapter have concentrated on plane-strain boundary-value problems for ideal fibre-reinforced materials, that is either incompressible materials containing one family of inextensible fibres or compressible materials reinforced by two orthogonal families of inextensible fibres. The governing equations for these problems are very simple because of the two geometrical constraints imposed on the deformation field. In this section we derive the basic equations for plane deformation of an elastic material reinforced by a single family of inextensible straight fibres. The analysis is based on papers by England, Ferrier and Thomas [11] and Morland [12].

4.1 Plane strain for compressible materials

Consider plane-strain deformations in the (x,y)-plane for a material which has inextensible fibres along the x-direction. Then $w \equiv 0$, $\partial u/\partial x = 0$ and u and v are independent of z so that

$$u = u(y). \tag{78}$$

It can no longer be concluded that $v = v(x)$ and, in fact, $v = v(x,y)$.

A compressible linearly elastic material reinforced by fibres along the x-direction has the constitutive equation

$$\sigma_{ij} = Ta_i a_j + \lambda e_{kk}\delta_{ij} + 2\mu_T e_{ij} + 2(\mu_L - \mu_T)(a_i a_k e_{kj} + a_j a_k e_{ik}),$$

see equation (24) of Chapter I. For plane-strain deformations with $\underset{\sim}{a} = (1,0,0)$ this becomes

$$\sigma_{xx} = \sigma_{11} = T + \lambda\frac{\partial v}{\partial y} = \bar{T},$$

$$\sigma_{yy} = \sigma_{22} = (\lambda+2\mu_T)\frac{\partial v}{\partial y},$$

$$\sigma_{zz} = \sigma_{33} = \lambda\frac{\partial v}{\partial y}, \tag{79}$$

$$\sigma_{xy} = \sigma_{12} = \mu_L\left\{u'(y) + \frac{\partial v}{\partial x}\right\},$$

$$\sigma_{xz} = \sigma_{yz} = 0.$$

Inserting these in the equilibrium equations gives

$$\frac{\partial \bar{T}}{\partial x} + \mu_L\left\{u''(y) + \frac{\partial^2 v}{\partial x \partial y}\right\} = 0,$$

$$\mu_L\frac{\partial^2 v}{\partial x^2} + (\lambda+2\mu_T)\frac{\partial^2 v}{\partial y^2} = 0.$$

Hence the displacement v normal to the fibres satisfies a Laplace-type equation. For convenience, if we introduce the constant

$$\alpha^2 = \mu_L/(\lambda+2\mu_T) \tag{80}$$

then v satisfies

$$\frac{\partial^2 v}{\partial x^2} + \frac{1}{\alpha^2}\frac{\partial^2 v}{\partial y^2} = 0. \tag{81}$$

The incompressible limit is attained as $\lambda \to \infty$, that is as $\alpha \to 0$.

The tension along the fibres integrates to give

$$\sigma_{xx} = \bar{T} = -\mu_L\left\{xu''(y) + \frac{\partial v}{\partial y}\right\} + F(y), \tag{82}$$

where F(y) is an arbitrary tension on each fibre, and the other in-plane stress components are

$$\sigma_{yy} = \frac{\mu}{\alpha^2}\frac{\partial v}{\partial y}, \qquad \sigma_{xy} = \mu\left\{u'(y) + \frac{\partial v}{\partial x}\right\}, \qquad \mu = \mu_L. \tag{83}$$

4.2 Generalised plane stress

It is straightforward to consider the in-plane deformation of a thin elastic plate reinforced by a single family of inextensible fibres. If the upper and lower faces of the plate are unstressed and the plate is deformed symmetrically about its mid-plane, then the mean displacements and stresses through the thickness of the plate satisfy the equations (78), (81) and (83), where α is given by the equation

$$\alpha^2 = \frac{\mu_L(\lambda+2\mu_T)}{(\lambda+2\mu_T)^2-\lambda^2} . \tag{84}$$

If the material is incompressible the equations have an identical form to those above except that $\alpha^2 = \mu_L/4\mu_T$. Thus the system of equations detailed above has a number of different physical interpretations. In the remainder of this section the material is referred to as a compressible material with a single family of fibres undergoing plane strain.

4.3 Boundary-value problems

The equations of anisotropic linear elasticity are elliptic in character. When we introduce the inextensibility constraint we change the nature of the system to one that is 'hyperbolic' in the x-direction and elliptic for the normal displacement v. The ideal material considered in the previous sections was 'hyperbolic' in both the x- and y-directions. For a body occupying a finite region it would seem reasonable that as the elastic moduli tend to infinity (or zero) the solution to the given problem would tend to that predicted by the one-fibre theory or the ideal theory. I am not certain that we can infer this when the body is infinite in extent. In fact, great care has to be taken in establishing and using the appropriate conditions at infinity for these plane problems (see England, Ferrier and Thomas [11]). For example, for the half-plane $y \geqslant 0$ (with polar angle satisfying $0 \leqslant \theta \leqslant \pi$) every fibre extends to infinity, whereas for the sector $\varepsilon \leqslant \theta \leqslant \pi-\varepsilon$ no fibres extend to infinity and the body can deform in a different manner. The following subsections are intended to illustrate the behaviour of the one-fibre compressible model by considering some common boundary-value problems.

4.4 Edge tractions on the half-plane $y > 0$

Consider the stress boundary-value problem

$$\sigma_{yy} = p(x)\ , \qquad \sigma_{xy} = s(x) \qquad \text{on} \qquad y = 0 \tag{85}$$

for the half-plane $y > 0$ where $p(x)$, $s(x)$ are sectionally continuous and have $O(|x|^{-k})$, $k > 1$ as $|x| \to \infty$. The fibres are parallel to the boundary $y = 0$.

The problem reduces to determining the function $v(x,y)$ which will satisfy the scaled Laplace equation (81) and the two boundary conditions

$$\frac{\mu}{\alpha^2}\frac{\partial v}{\partial y} = p(x)\ , \qquad \mu\left\{u'(0) + \frac{\partial v}{\partial x}\right\} = s(x) \qquad \text{on} \qquad y = 0\ . \tag{86}$$

Clearly we can choose v to satisfy only one of the boundary conditions (86). If we choose to satisfy $(86)_2$ the solution would lead to discontinuous values of σ_{yy} across the fibre $y = 0$ which could not be equilibrated. Consequently we must satisfy $(86)_1$ and allow discontinuous values of σ_{xy} across the edge fibre which can be equilibrated by a finite force along the fibre.

The solution to this problem may be obtained by complex variable or Fourier transform methods. On Fourier transforming (81) we find

$$\bar{v} = \int_{-\infty}^{\infty} e^{i\xi x} v(x,y)\ dx = A(\xi)e^{-\alpha|\xi|y}$$

and from $(86)_1$

$$A(\xi) = -\frac{\alpha}{\mu}\frac{\bar{p}(\xi)}{|\xi|}\ e^{-\alpha|\xi|y}\ .$$

On inverting we find

$$\frac{\partial v}{\partial y} = \frac{\alpha^2}{\mu}\int_{-\infty}^{\infty} p(x-t)G(t,y)\ dt \tag{87}$$

where

$$G(t,y) = \frac{\alpha}{\pi}\frac{y}{t^2+\alpha^2 y^2} . \tag{88}$$

Hence

$$\frac{\partial v}{\partial y} = \frac{\alpha}{\pi\mu y}\int_{-\infty}^{\infty} p(t)\, dt + O\left(\frac{1}{y^3}\right) \qquad \text{for large } |y| .$$

If the resultant normal force is zero, then v is bounded at infinity otherwise (as in linear elasticity) v has a logarithmic singularity at infinity. The fibres carry the stress

$$\sigma_{xx} = -\mu\left\{xu''(y) + \frac{\partial v}{\partial y}\right\} + F(y)$$

and since $\partial v/\partial y$ is bounded at infinity, σ_{xx} is bounded at infinity only if

$$u''(y) = 0 .$$

Then $u(y) = A + By$ and if the stress field is zero at infinity then $F(y) \equiv 0$ and, from the form of σ_{xy}, $B = 0$. Hence $\sigma_{xx} = -\alpha^2\sigma_{yy}$ for $y > 0$.

Note, however, that the shear stress is discontinuous across $y = 0$ so that

$$\sigma_{xy} = \begin{cases} \mu\partial v/\partial x & \text{on } y = 0+ , \\ s(x) & \text{on } y = 0- . \end{cases}$$

This can only be equilibrated if the edge fibre carries the tension

$$T(x) = -\mu v(x,0) + \int_0^x s(x)\, dx + T_0 ,$$

where T_0 is an arbitrary constant.

As a particular example let us suppose

$$p(x) = \begin{cases} -p & \text{for } |x| \leqslant a , \\ 0 & \text{for } |x| > a , \end{cases} \qquad s(x) \equiv 0 . \tag{89}$$

Then we find

$$\sigma_{yy} = \frac{\mu}{\alpha^2}\frac{\partial v}{\partial y} = -\frac{p}{\pi}\left\{\arctan\left(\frac{a-x}{\alpha y}\right) + \arctan\left(\frac{a+x}{\alpha y}\right)\right\} .$$

As in classical elasticity this has a simple interpretation in terms of angles (see Figure 15)

$$\sigma_{yy} = -\frac{p}{\pi}(\phi_1+\phi_2)\ .$$

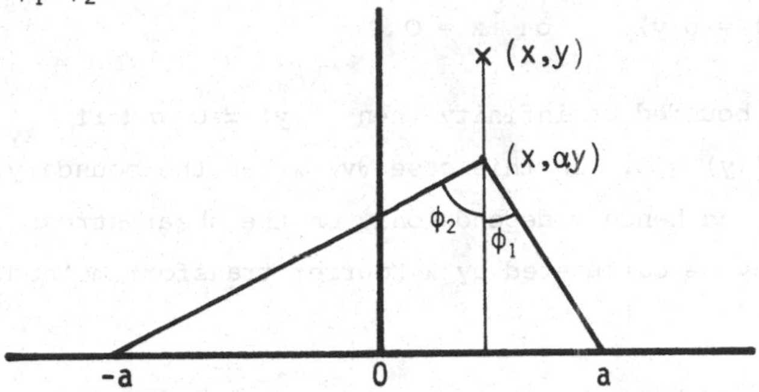

Figure 15

The shear stress across y = 0 is discontinuous and, after some manipulation, we find

$$\sigma_{xy} = \frac{\alpha p}{\pi}\log\left|\frac{a-x}{a+x}\right| \qquad \text{on} \quad y = 0+$$

so that along the edge fibre

$$T(x) = -\mu v(x,0) = \frac{\alpha p}{\pi}\{(a-x)\log|a-x|+(a+x)\log|a+x|-2a\} \tag{90}$$

to within an additive constant. Hence the force along the edge fibre is finite for all finite x but displays a logarithmic singularity as $|x| \to \infty$. Had we chosen a problem in which the resultant force applied to the half-space was zero, the force in the edge fibre would have been bounded at infinity.

4.5 Edge tractions on the half-plane x > 0

In this case the fibres are normal to the edge x = 0. Suppose the boundary conditions are

$$\sigma_{xx} = p(y)\ , \qquad \sigma_{xy} = s(y) \qquad \text{on} \quad x = 0\ , \tag{91}$$

where p(y) and s(y) are sectionally continuous and have $O(|y|^{-k})$, $k > 1$, as $|y| \to \infty$. Now v satisfies the scaled Laplace's equation in x > 0 subject to the boundary conditions

$$u'(y) + \frac{\partial v}{\partial x} = \frac{1}{\mu} s(y) \qquad \text{on} \quad x = 0 ,$$

$$- \mu \frac{\partial v}{\partial y} + F(y) = p(y) \qquad \text{on} \quad x = 0 . \tag{92}$$

If σ_{xx} is to be bounded at infinity then $u''(y) \equiv 0$ and if σ_{xy} is zero at infinity then $u'(y) \equiv 0$. In this case $\partial v/\partial x$ has the boundary value $s(y)/\mu$ on $x = 0$ and hence v depends only on the shear stress on the boundary. It may be calculated by a Fourier transform method. Given v, then

$$F(y) = p(y) + \mu \frac{\partial v}{\partial y}(0,y)$$

and hence $p(y)$ only influences the tension in the fibres; in fact

$$\sigma_{xx} = p(y) + \mu \frac{\partial v}{\partial y}(0,y) - \mu \frac{\partial v}{\partial y}(x,y) .$$

We now look at two particular cases:

Case 1

$$p(y) = \begin{cases} -p_0 , & |y| < a , \\ 0 , & |y| > a , \end{cases} \qquad s(y) \equiv 0 .$$

In this case $v \equiv 0$ and the solution is simply a band of pressure transmitted along the fibres lying in the region $|y| \leqslant a$. The rest of the half-plane is unstressed.

Case 2

$$p(y) \equiv 0 , \qquad s(y) = \begin{cases} s , & |y| < a , \\ 0 , & |y| > a . \end{cases}$$

Then, from (92),

$$\frac{\partial v}{\partial x} = \frac{s}{\pi\mu}\left[\arctan\left\{\frac{\alpha(a-y)}{x}\right\} + \arctan\left\{\frac{\alpha(a+y)}{x}\right\}\right] ,$$

and

$$\frac{\partial v}{\partial y} = \frac{s\alpha}{\pi\mu}\left[2 + \tfrac{1}{2}\log\left\{\frac{x^2+\alpha^2(a+y)^2}{x^2+\alpha^2(a-y)^2}\right\}\right] ,$$

so that

$$\sigma_{xx} = \frac{s\alpha}{2\pi}\log\left\{\left(\frac{a+y}{a-y}\right)^2 \frac{x^2+\alpha^2(a-y)^2}{x^2+\alpha^2(a+y)^2}\right\}.$$

Thus the fibres $y = \pm a$ carry infinite stresses to infinity. The stresses elsewhere in the half-plane are finite.

4.6 Infinite strip

To illustrate this type of problem consider the infinite strip $|y| \leqslant h$ reinforced by inextensible fibres parallel to the faces $y = \pm h$. Suppose the stress distributions

$$\begin{aligned} \sigma_{yy} &= p^+(x), \quad \sigma_{xy} = s^+(x) \quad \text{on } y = h, \\ \sigma_{yy} &= p^-(x), \quad \sigma_{xy} = s^-(x) \quad \text{on } y = -h, \end{aligned} \tag{93}$$

are applied to the strip. Let us also suppose, for simplicity, that the problem is symmetrical about $x = 0$ so that $u(y) = 0$, and that the applied stress distributions have a zero resultant force. The normal displacement v may be used to equilibrate the normal stress σ_{yy} on $y = \pm h$ by solving the boundary value problem

$$\begin{aligned} \frac{\partial^2 v}{\partial x^2} + \frac{1}{\alpha^2}\frac{\partial^2 v}{\partial y^2} &= 0 \quad \text{in } -h < y < h, \\ \frac{\partial v}{\partial y} &= \frac{\alpha^2}{\mu} p^{\pm}(x) \quad \text{on } y = \pm h. \end{aligned} \tag{94}$$

The shear stress conditions on $y = \pm h$ cannot be satisfied and consequently the edge fibres must be singular fibres and carry the forces $T^+(x)$ and $T^-(x)$. It is simple to show that

$$\begin{aligned} T^+(x) &= \mu v(x,h) - \int_0^x s^+(t)\,dt, \\ T^-(x) &= -\mu v(x,-h) + \int_0^x s^-(t)\,dt, \end{aligned} \tag{95}$$

to within arbitrary constants. The shear stress boundary values only influence the solution through the forces in the edge fibres and so the terms $s^{\pm}(x)$ will be ignored in the following discussion.

The boundary-value problem (94) may be solved by a Fourier transform technique to give

$$\frac{\mu}{\alpha^2}\frac{\partial\bar{v}}{\partial y} = \bar{p}^+(\xi)\,\frac{\sinh\{\xi\alpha(h+y)\}}{\sinh(2\xi\alpha h)} + \bar{p}^-(\xi)\,\frac{\sinh\{\xi\alpha(h-y)\}}{\sinh(2\xi\alpha h)}$$

and this may be inverted and $v(x,y)$ written as a convolution integral (see England, Ferrier and Thomas [11]). It is straightforward to show that as $\alpha \to 0$ the solution obtained tends to that predicted by the ideal theory.

Let us consider the three-point bending problem for which

$$p^+(x) = -2p\delta(x)\,, \qquad p^-(x) = -p\{\delta(x-a) + \delta(x+a)\}\,. \tag{96}$$

We see at once that the presence of point forces will give rise to logarithmic singularities in $v(x,y)$ at the points of application of these forces. In addition, since the tension in each edge fibre is proportional to its normal displacement (see (95)), these tensions will have logarithmic singularities under the point forces. After some manipulation we find

$$T^+(x) = \frac{p\alpha}{\pi}\log\left[\frac{\sinh^2(\pi x/4\alpha h)}{\cosh\{\pi(x-a)/4\alpha h\}\cosh\{\pi(x+a)/4\alpha h\}}\right],$$

$$T^-(x) = -\frac{p\alpha}{\pi}\log\left[\frac{\cosh^2(\pi x/4\alpha h)}{\sinh(\pi|x-a|/4\alpha h)\sinh(\pi|x+a|/4\alpha h)}\right], \tag{97}$$

where the arbitrary constants are chosen so that these tensions tend to zero at infinity. Note that $T^+(x)$ is compressive and is singular at the point of application $(x = 0)$ of the point force.

The tension $T^-(x)$ on the lower edge is compressive in $|x| > a$ and is mainly compressive in $|x| < a$. It will be tensile near $x = 0$ only if

$$\frac{\pi a}{4\alpha h} > \operatorname{arcsinh} 1 = 0.8814\,. \tag{98}$$

This result indicates that if the depth of the beam is sufficiently great then both edge fibres will be under compression or, as an alternative, only if the spacing of the point forces is sufficiently large will the lower fibre come under tension. It is apparent that, for a given a and h, as the constant α tends to zero a tensile state will prevail. In this case there is a tensile force over the range

$$|x| < \frac{2\alpha h}{\pi} \operatorname{arccosh}\left\{\sinh^2\left(\frac{\pi a}{4\alpha h}\right)\right\}. \tag{99}$$

The normal displacement on the surface is proportional to $T^{\pm}(x)$ and is finite at all interior points. The stress field $\sigma_{xx} = -\alpha^2\sigma_{yy}$ and σ_{xy} is bounded at all interior points of the strip and tends to zero at infinity.

The behaviour of the system as $\alpha \to 0$ (as the material becomes incompressible) tends to the solution for an ideal material. Figure 16 gives the value of the tension in the lower edge fibre for a beam with $a = 1$, $h = 1$ for a range of values of the constant $\alpha = \{\mu_T/(\lambda+2\mu_T)\}^{\frac{1}{2}}$. Note that when $\alpha = 1$ the force is entirely compressive, but for small values of α it is tensile almost everywhere in $|x| \leqslant 1$ and close to the solution for the ideal material at all points except in a small neighbourhood near the point force at $x = 1$.

J. N. Thomas [13] has performed some finite element calculations for an orthotropic material in which various moduli were taken smaller progressively and the one-fibre material and the ideal material behaviour obtained as a limit. If we adopt the notation for plane strain (or plane stress) of an orthotropic material

$$e_{xx} = A\sigma_{xx} - B'\sigma_{yy}, \qquad e_{yy} = -B'\sigma_{xx} + C\sigma_{yy}, \qquad e_{xy} = \sigma_{xy}/2\mu \tag{100}$$

and introduce the Airy stress function by

$$\sigma_{xx} = \chi_{,yy}, \qquad \sigma_{yy} = \chi_{,xx}, \qquad \sigma_{xy} = -\chi_{,xy} \tag{101}$$

then the strain compatibility equation becomes

$$A\chi_{,xxxx} + B\chi_{,xxyy} + C\chi_{,yyyy} = 0$$

where $B = (1/\mu) - 2B'$. We can factorise this to give

$$\left(\varepsilon^2 \frac{\partial^2}{\partial x^2} + \frac{\partial^2}{\partial y^2}\right)\left(\frac{\partial^2}{\partial x^2} + c^2 \frac{\partial^2}{\partial y^2}\right)\chi = 0. \tag{102}$$

For the isotropic case $\varepsilon = c = 1$.

For a one-fibre strong material

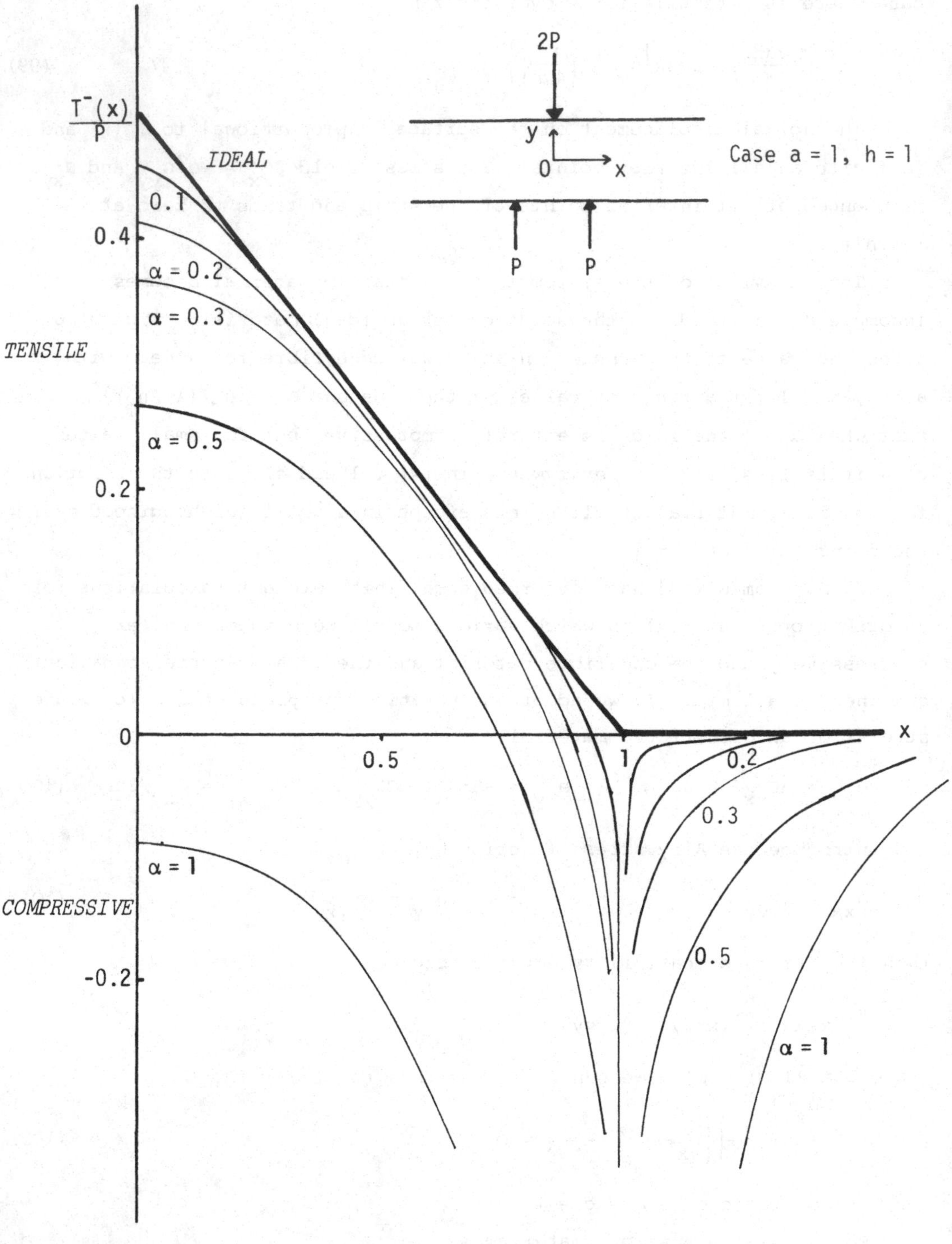

Figure 16. One-fibre compressible material: force along the edge fibre y = -1

$$\frac{\mu}{E_1} \ll 1 , \qquad \frac{E_2}{E_1} \ll 1 , \qquad \nu_{23} \simeq 1 ,$$

so that

$$\varepsilon^2 \simeq \frac{\mu}{E_1} , \qquad c^2 \simeq (1-\nu_{23})\frac{2\mu}{E_2} .$$

The constant α used above in the inextensible theory is approximately equal to c. For the two-fibre case $\varepsilon^2 = c^2 \simeq \mu/E_1$.

Special values of these moduli are

Graphite-epoxy (Fahmy [14]), $\varepsilon = 0.218$, $c = 0.738$,

Carbon-epoxy (Markham [15]), $\varepsilon = 0.156$, $c = 0.747$.

The following figures give some of Thomas's results for the three-point bending problem for a beam with dimensions 10×100 with point forces applied ± 10 units from the centre line. The origin is taken at the centre of the beam. The normal displacement of the centre line on $y = 0$ is given in Figure 17. The remaining figures compare the force acting across the boundary layer with that predicted by the ideal theory. In particular, Figure 18 compares the force carried by the edge fibre $y = 5$ (as predicted by the ideal theory) with the resultant force

$$\int_{5-d}^{5} \sigma_{xx}\, dy$$

over a boundary layer of thickness d (where d takes the values 0.5, 1 and 2) in an orthotropic elastic material with various elastic moduli. The continuous line indicates the force over the estimated thickness of the boundary layer $d = 10\varepsilon$ (see Chapter IV). Figure 19 makes the same comparison for the lower edge fibre $y = -5$ and Figures 20 and 21 compare the forces carried by strips of width 2d ($d = 0.5, 1, 2$) about the lines $x = 0$ and $x = 10$ (i.e. below the point forces) with the predictions of the ideal theory. It will be seen that except in the neighbourhood of the point forces good agreement is obtained between the numerical results for strong orthotropic elastic materials and the ideal theory. Note, for convenience, Thomas took the magnitude of the point force p to be 2×10^7N/m which gives rise to a factor 2 throughout Figures 18 to 21.

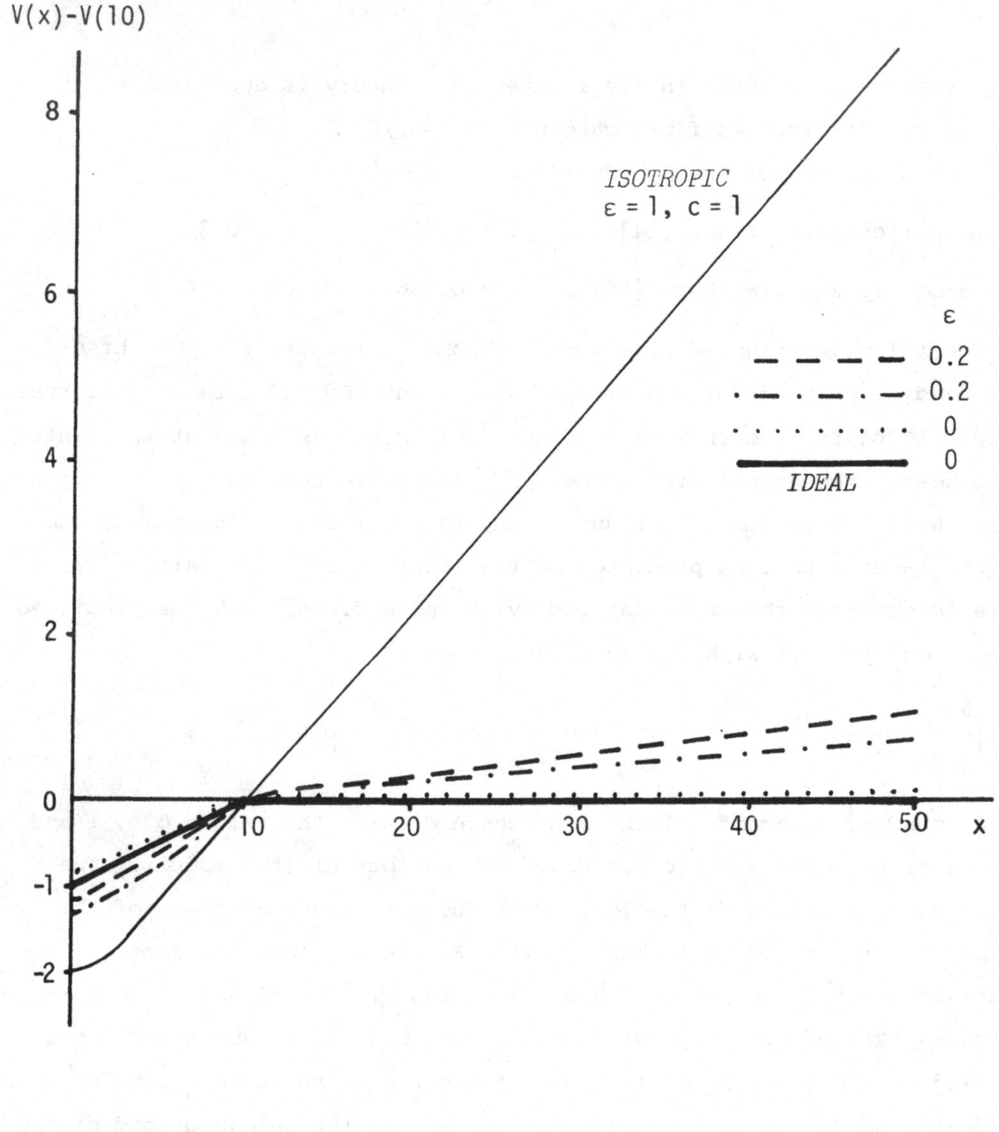

Figure 17. Normal displacement of the centreline in a 3-point bending test.
$V(x) = (\mu/p)\,v(x,0)$

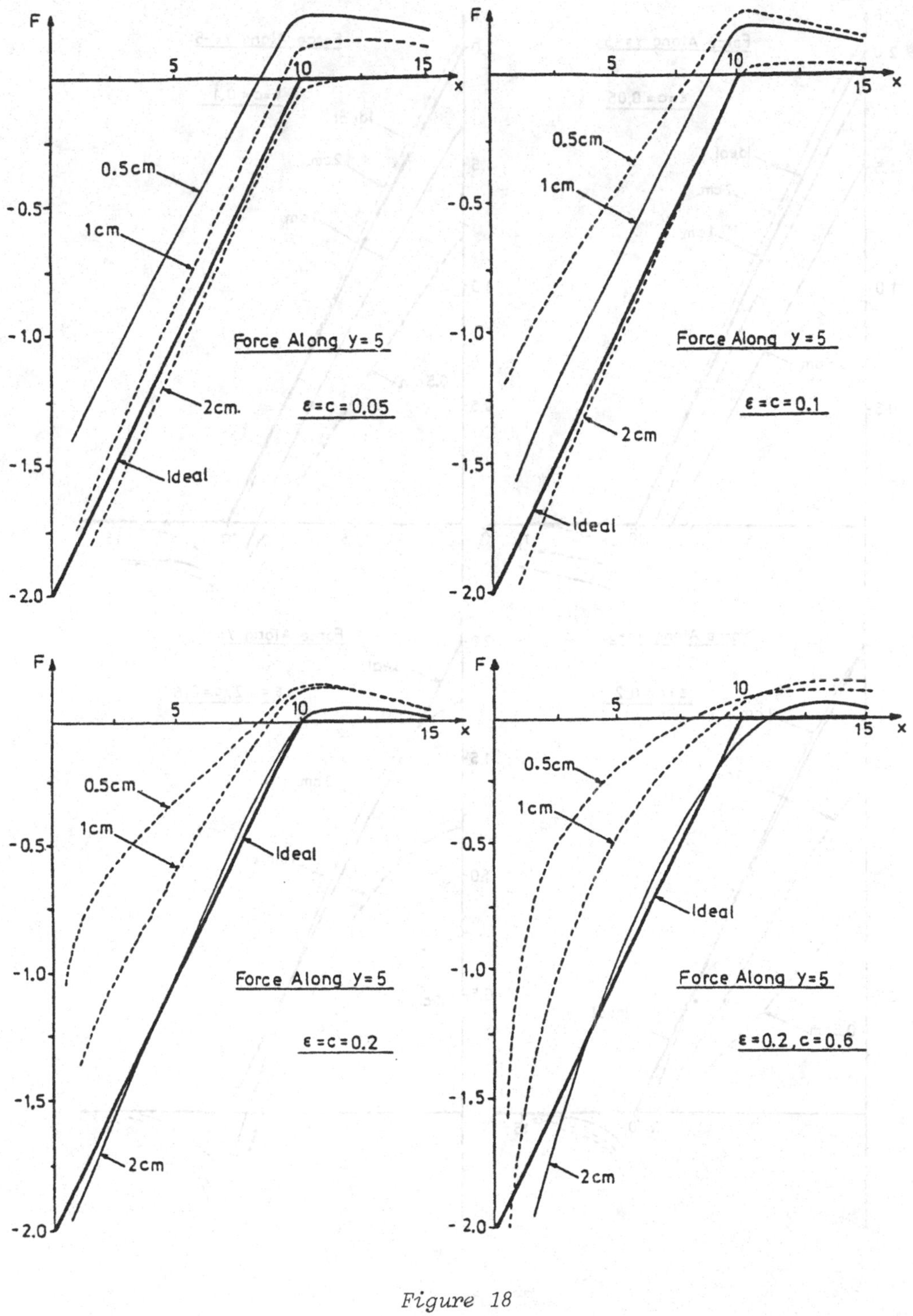
F
5
10
15
x
0.5cm
1cm
2cm.
Ideal
Force Along y= 5
ε=c=0.05
-0.5
-1.0
-1.5
-2.0
F
5
10
15
x
0.5cm
1cm
2cm
Ideal
Force Along y=5
ε=c=0.1
-0.5
-1.0
-1.5
-2.0
F
5
10
15
x
0.5cm
1cm
Ideal
2cm
Force Along y=5
ε=c=0.2
-0.5
-1.0
-1.5
-2.0
F
5
10
15
x
0.5cm
1cm
Ideal
2cm
Force Along y=5
ε=0.2, c=0.6
-0.5
-1.0
-1.5
-2.0

Figure 18

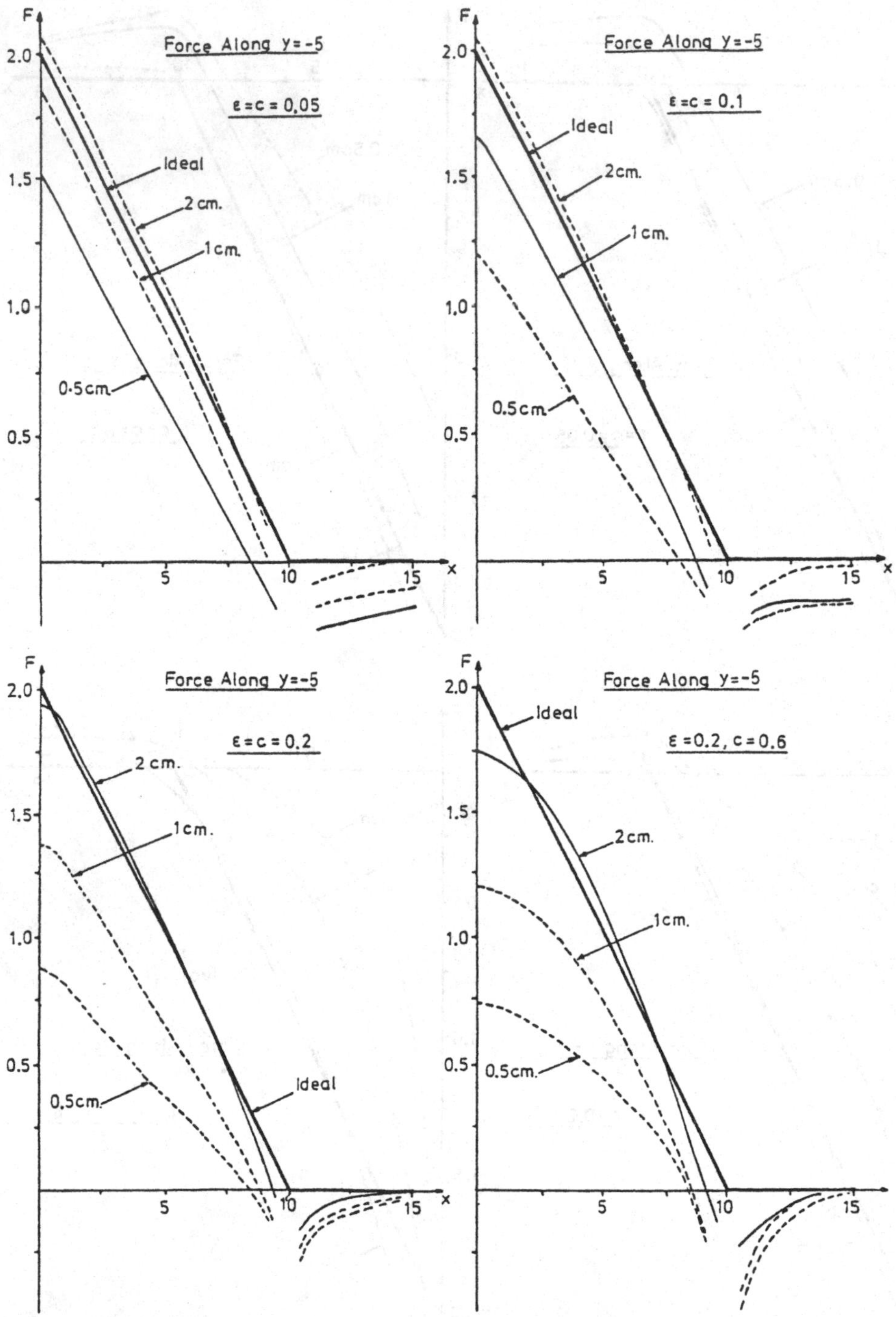
Force Along y=-5
ε=c=0,05
Ideal
2 cm.
1cm.
0·5cm
Force Along y=-5
ε=c=0.1
Ideal
2cm.
1cm.
0.5cm
Force Along y=-5
ε=c=0.2
2 cm.
1cm.
0,5cm.
Ideal
Force Along y=-5
ε=0.2, c=0.6
Ideal
2cm.
1cm.
0.5cm.
F
x
2.0
1.5
1.0
0.5
5
10
15

Figure 19

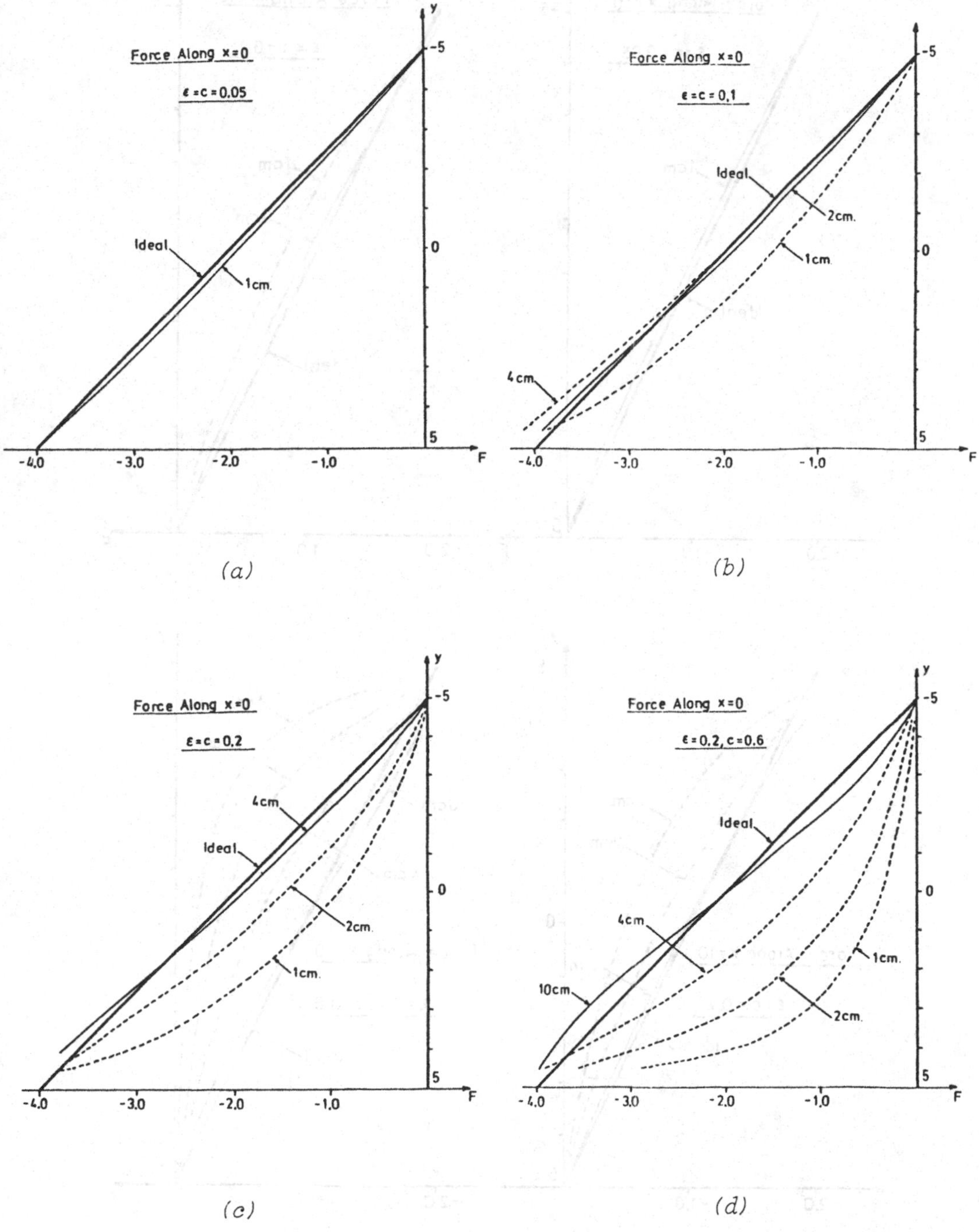
Force Along x=0
ε=c=0.05
Ideal
1cm.
y
-5
0
5
F
-4.0
-3.0
-2.0
-1.0
(a)
Force Along x=0
ε=c=0.1
Ideal
2cm.
1cm.
4cm
(b)
Force Along x=0
ε=c=0.2
4cm
Ideal
2cm.
1cm.
(c)
Force Along x=0
ε=0.2, c=0.6
Ideal
4cm
10cm
1cm.
2cm.
(d)

Figure 20

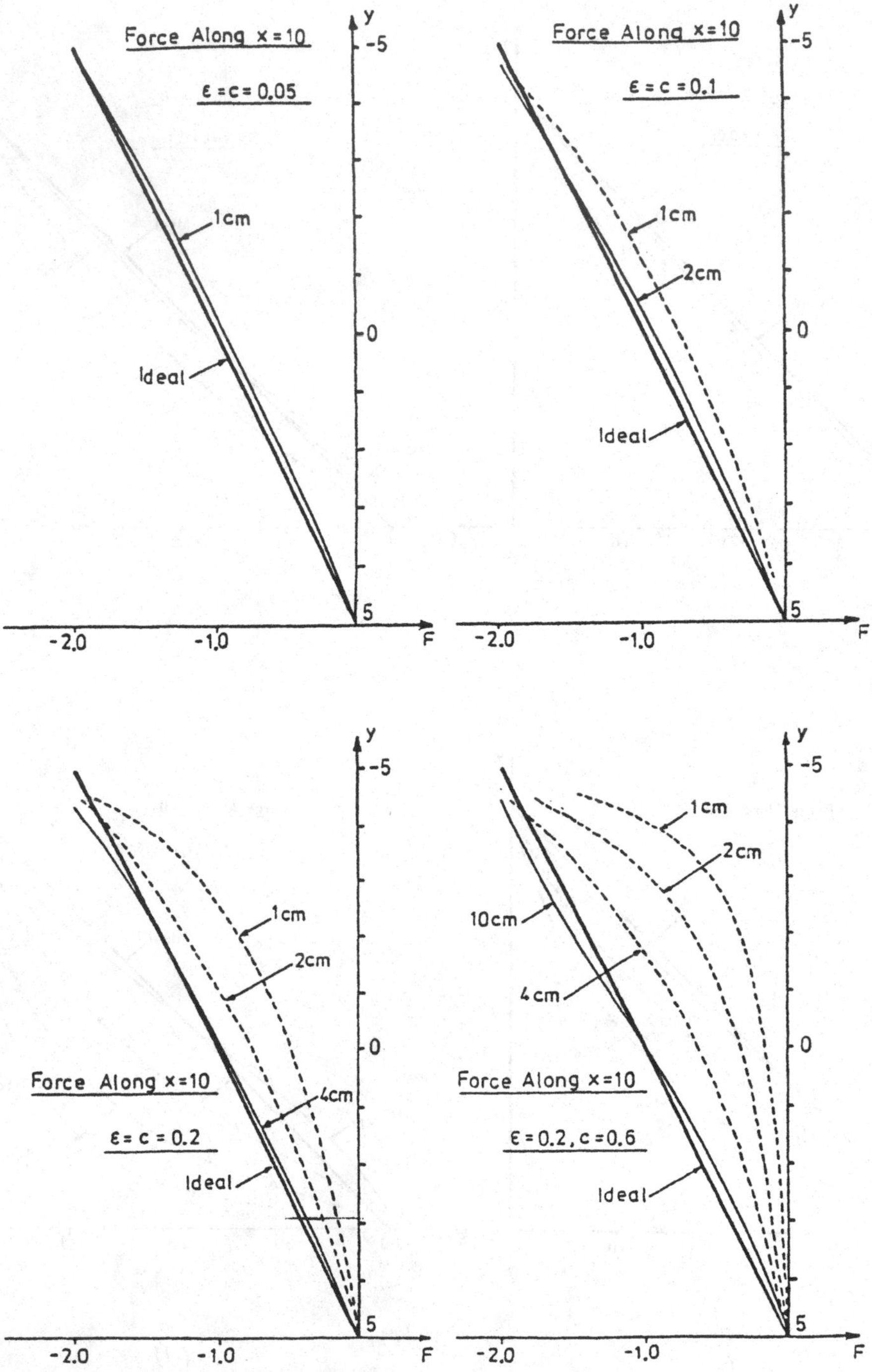
Force Along x = 10
ε = c = 0.05
1cm
Ideal
-5
0
5
y
F
-2.0
-1.0
Force Along x = 10
ε = c = 0.1
1cm
2cm
Ideal
Force Along x = 10
ε = c = 0.2
1cm
2cm
4cm
Ideal
Force Along x = 10
ε = 0.2, c = 0.6
1cm
2cm
4cm
10cm
Ideal

Figure 21

4.7 Finite regions

The solution of a boundary-value problem for a finite region composed of this one-fibre material reduces to the solution of the scaled Laplace's equation in the region. In general this is a numerical problem but for certain shapes analytical methods may be applied. Consider the problem illustrated in Figure 22.

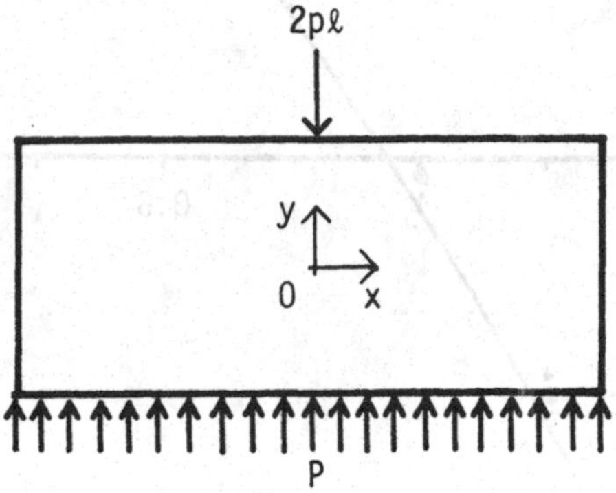

Figure 22

Take the origin at the centre of the rectangle and suppose the fibres are parallel to the x-axis. Then, by symmetry, $u(y) \equiv 0$ and the problem reduces to solving the scaled Laplace equation (81) in the rectangle subject to boundary conditions

$$\sigma_{xy} = 0 \Rightarrow \frac{\partial v}{\partial x} = 0 \qquad \text{on} \quad x = \pm\ell ,$$

$$\frac{\partial v}{\partial y} = \frac{\alpha^2}{\mu} p^{\pm}(x) \qquad \text{on} \quad y = \pm h ,$$

using an obvious notation. (Note the boundary condition $\sigma_{xy} = 0$ on $y = \pm h$ has not been included in the formulation so far.) When $p^{+}(x) = -2p\ell\delta(x)$ and $p^{-}(x) = p$, $|x| \leqslant \ell$, this problem has the solution

$$v = -\frac{p\alpha^2}{\mu} y - \frac{p\alpha}{\mu} \sum_{1}^{\infty} \frac{2\ell}{n\pi} \cos\left(\frac{n\pi x}{\ell}\right) \frac{\cosh\{n\pi\alpha(y+h)/\ell\}}{\sinh(2n\pi\alpha h/\ell)} \qquad (103)$$

to within a rigid body displacement. The solution to the same problem for an ideal material is

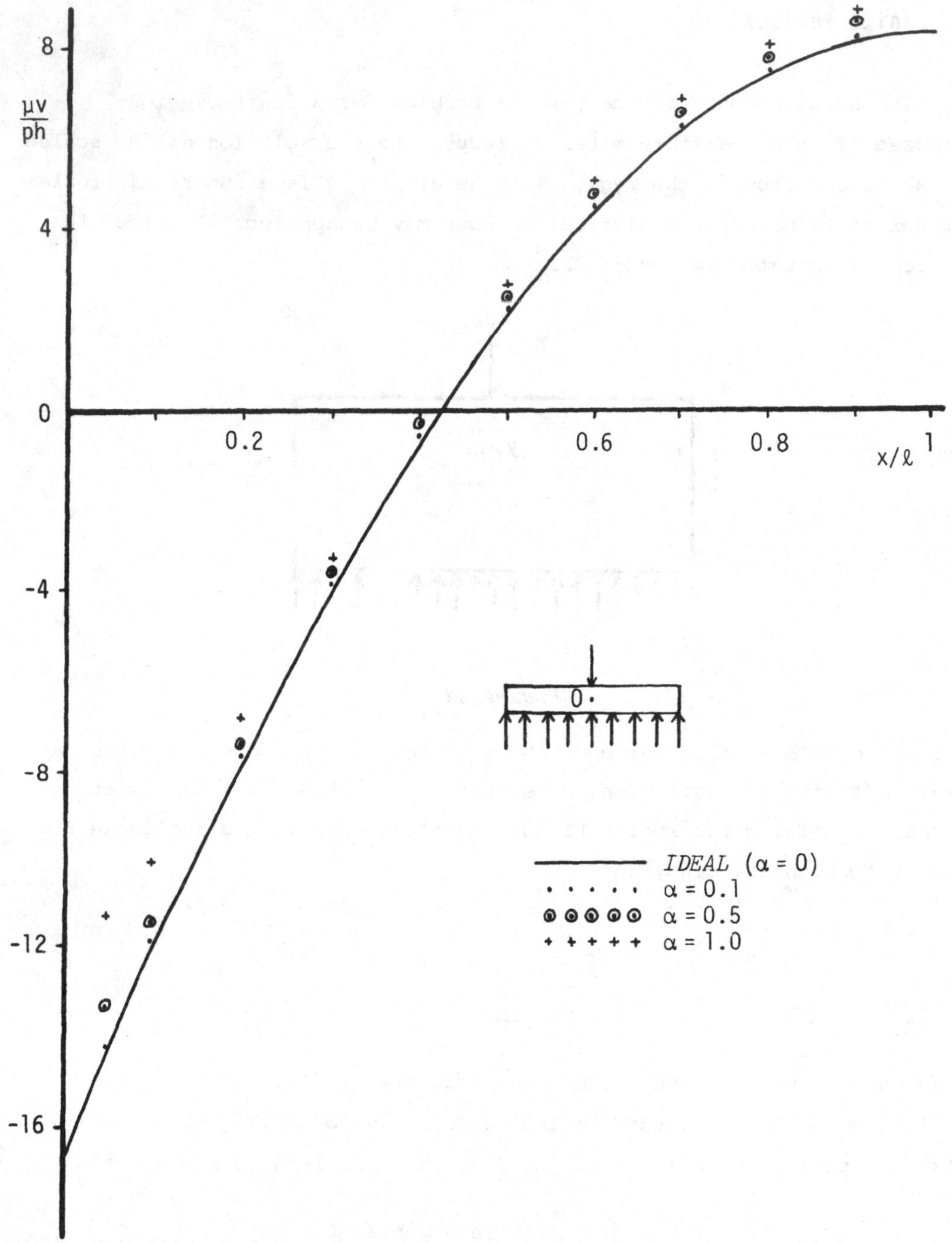

Figure 23. Bending of a finite beam: scaled normal deflection μv/ph on y = -h

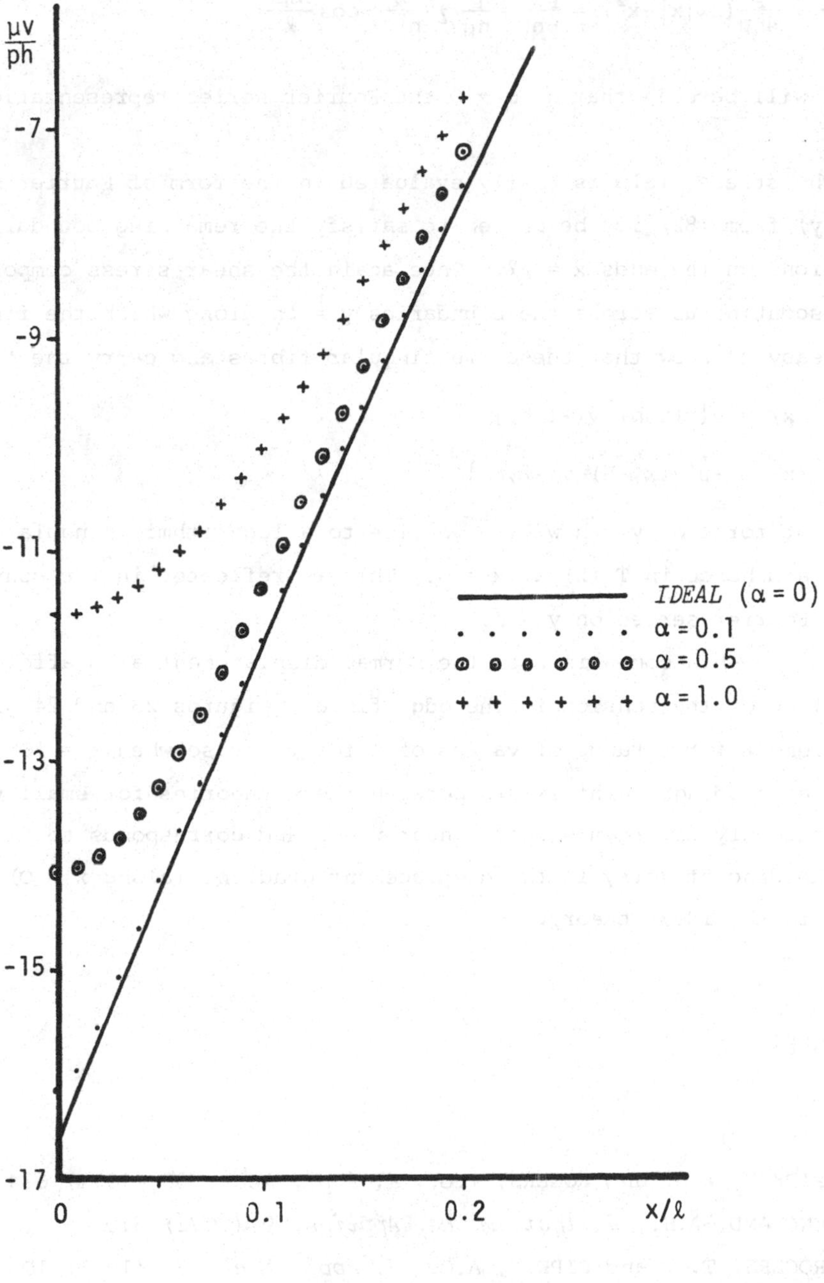

Figure 24. Bending of a finite beam: scaled normal deflection on y = -h *near* x = 0

$$v = \frac{p}{4h\mu}(2\ell|x|-x^2) = \frac{p\ell^2}{6h\mu} - \frac{p}{h\mu}\sum_1^\infty \frac{\ell^2}{n^2\pi^2}\cos\frac{n\pi x}{\ell},$$

and it will be seen that as $\alpha \to 0$ the Fourier series representations agree.

The stress field is easily evaluated in the form of Fourier series and $F(y)$ from (82) may be chosen to satisfy the remaining boundary conditions on the ends $x = \pm\ell$. Once again the shear stress components are discontinuous across the boundaries $y = \pm h$ along which the fibres lie. It is easy to show that these are singular fibres and carry the tensions

$$T^+(x) = \mu\{v(x,h)-v(-\ell,h)\},$$

$$T^-(x) = -\mu\{v(x,-h)-v(-\ell,h)\}.$$

The point force on $y = h$ will give rise to a logarithmic singularity in $v(x,h)$ and hence in $T^+(x)$ at $x = 0$. This is reflected in the convergence of the Fourier series on $y = h$.

On $y = -h$ a comparison of the normal displacement also affords a comparison of the tension in the edge fibre. Figures 23 and 24 give this displacement for a range of values of α for the case when $\ell = 10h$. It is seen that good agreement exists between these theories for small values of α; the only divergence occurs near $x = 0$ and corresponds to "ironing out" the discontinuity in the displacement gradient (along $x = 0$) which occurs in the ideal theory.

REFERENCES

[1] PIPKIN, A.C. and ROGERS, T.G., *J. Appl. Mech.* 38 (1971) 634

[2] ENGLAND, A.H., *J. Inst. Maths. Applics.* 9 (1972) 310

[3] ROGERS, T.G. and PIPKIN, A.C., *J. Appl. Mech.* 38 (1971) 1047

[4] ENGLAND, A.H. and ROGERS, T.G., *Q. Jl. Mech. Appl. Math.* 26 (1973) 303

[5] ENGLAND, A.H. and THOMAS, J.N., *J. Inst. Maths. Applics.* 14 (1974) 347

[6] THOMAS, J.N., *Z. Angew. Math. Phys.* 26 (1974) 553

[7] BOWIE, O.L. and FREESE, C.E., *Int. J. Fracture Mech.* 8 (1972) 49

[8] BISHOP, S.M., *Royal Aircraft Establishment Tech. Reports* 72026 and 73124 (1973)

[9] EVERSTINE, G.C. and PIPKIN, A.C., *J. Appl. Mech.* 40 (1973) 518

[10] SPENCER, A.J.M., *Int. J. Sol. Struct.* 10 (1974) 1103

[11] ENGLAND, A.H., FERRIER, J.E. and THOMAS, J.N., *J. Mech. Phys. Solids* 21 (1973) 279

[12] MORLAND, L.W., *Int. J. Sol. Struct.* 9 (1973) 1501

[13] THOMAS, J.N., *Ph.D. Thesis, University of Nottingham* (1974)

[14] FAHMY, A.A. and RAGAI-ELLOZY, A.N., *Jl. Comp. Materials* 8 (1974) 90

[15] MARKHAM, M.F., *Composites* 1 (1970) 145

IV

STRESS CHANNELLING AND BOUNDARY LAYERS IN STRONGLY ANISOTROPIC SOLIDS

A. C. PIPKIN

Division of Applied Mathematics
Brown University
Providence, Rhode Island 02912
U.S.A.

1. Introduction

Problems involving materials reinforced with strong fibers can be simplified very greatly by introducing the approximation that the fibers are inextensible. This leads to some striking results that at first do not appear to be correct, but which on closer examination are seen to be clear and forceful representations of real phenomena, phenomena which occur in highly anisotropic materials but not in the more familiar isotropic materials.

As a quick example [1], consider plane strain of a slab bonded to a rigid support at one end and subjected to a uniform shearing traction at the other, so that the displacement components u and v satisfy $u(0,y) = v(0,y) = 0$ at the fixed end $x = 0$ and the shearing stress is $\sigma_{xy}(L,y) = \sigma$ at the end $x = L$. If the material is reinforced with strong fibers that run in the x-direction, as an idealization we can treat these fibers as inextensible, and treat the composite material as a homogeneous continuum. Then u is constant along each fiber y, and since u(y) is zero at $x = 0$, it follows that $u = 0$ everywhere.

The result, from inextensibility, that $u = u(y)$, means that Saint-Venant's principle cannot be used in the usual way. The principle says in effect that except close to the end $x = 0$, it does not matter whether the boundary conditions at $x = 0$ are satisfied or not. In highly anisotropic materials this is not the case. Disburbances can be transmitted along fibers for a long distance, so "end effects" may be the most prominent part of the solution.

Returning to the problem, let us suppose that the stress response of the material is linearly elastic, apart from the tensile stress σ_{xx} that arises as a reaction to the constraint of inextensibility. Then with $u = 0$, we have $\sigma_{xy} = Gv,_x$ and $\sigma_{yy} = E'v,_y$, and the equilibrium equation implies that v satisfies a generalized Laplace equation. The boundary conditions on v come from the conditions that $\sigma_{xy} = \sigma$ at $x = L$ and that $\sigma_{yy} = 0$ on $y = 0$ and $y = H$, and that $v = 0$ at $x = 0$. The unique solution satisfying these conditions is $v = (\sigma/G)x$.

The boundaries $y = 0$ and $y = H$ are supposed to be traction-free, so the specified shearing stress, $\sigma_{xy} = 0$, does not match the interior shearing stress $\sigma_{xy} = \sigma$. Accepting this, the shearing stress is discontinuous at the boundaries. However, with no stress on the bottom side of the fiber $y = 0$ and a uniform shear σ on its top side, elementary mechanics shows that there must be a finite tension $F(x) = \sigma(L-x)$ in the fiber $y = 0$. In terms of stress, this is represented as $\sigma_{xx} = F(x)\delta(y)$, where $\delta(y)$ is a Dirac delta. Since the fiber $y = 0$ is postulated to be inextensible, even an infinite stress in it does not cause it to deform, so there is no objection to this kind of singularity.

Singular fibers such as the fiber $y = 0$ in this example occur very commonly in solutions based on the assumption of inextensibility. The present lectures are devoted mainly to explaining and interpreting such singularities. As our basic model we use the theory of elastic deformations of materials that have a tensile modulus E for stretching in the x-direction that is very large in comparison to the shear modulus G and the modulus E' for tension in the y-direction. We show that whenever the theory of inextensible materials yields a singular fiber, elasticity theory gives a long, thin region in which the tensile stress σ_{xx} is very high but finite.

The mathematical theory is based on the smallness of a parameter ε that is approximately

$$\varepsilon \cong \sqrt{G/E}. \tag{1.1}$$

If L is a characteristic length defined by the boundary conditions of a problem, the thickness of a stress concentration layer is typically of the order of εL, and the tensile stress in it is of the order of $F(x)/\varepsilon L$, where $F(x)$ is the singular fiber force given by the inextensible theory. In cases in which the inextensible theory predicts unattenuated transmission of stress infinitely far along a fiber, elasticity theory shows that the actual attenuation length is of the order of L/ε.

For our present purpose we do not need to give separate consideration to materials composed of two orthogonal families of strong fibers. For such materials the ratio G/E' is also small, and a parameter c defined approximately by

$$c \cong \sqrt{G/E'} \tag{1.2}$$

arises from the theory. The same kind of analysis that we outline in these lectures can be used to show that a stress concentration layer in the y-direction will have a thickness of the order of cL, and that the attenuation length for stress transmission in the y-direction is of the order of L/c.

After outlining the theory of elastic plane stress, by detailed examination of a particular example we show how the result of the inextensible theory can be viewed as a limit of the elastic solution, including the appearance of a singular fiber. A different limiting process gives the details of the stress in the stress concentration layer. In the terminology of singular perturbation theory, the two kinds of limits are "outer" and "inner" limits. Guided by the inner limiting process in the particular example considered, we formulate a theory of stress concentration layers by applying the inner limiting process to a more general class of problems.

Since our main object is to learn how to interpret the results of the inextensible theory in terms of elastic solutions, we pay some attention to cases in which there can be an infinite stress singularity inside a stress concentration layer. Such singularities are not revealed by the inextensible theory, but their locations are easy to guess and their strengths are easy to compute. In particular, we show how to compute the elastic stress intensity factor at a crack tip from the force in the singular fiber passing through the tip, as computed from the inextensible theory.

2. Plane deformations of highly anisotropic materials [2].

For infinitesimal elastic deformations of transversely isotropic or orthotropic materials, the stress-strain relations for (generalized) plane stress have the form

$$E\varepsilon_{xx} = \sigma_{xx}-\nu\sigma_{yy} ,$$
$$E'\varepsilon_{yy} = \sigma_{yy}-\nu'\sigma_{xx} , \qquad (2.1)$$
$$2G\varepsilon_{xy} = \sigma_{xy} ,$$

where

$$\nu'/E' = \nu/E . \qquad (2.2)$$

As in any plane problem, the stress equations of equilibrium are satisfied if the stress components are defined in terms of Airy's stress potential χ by

$$\sigma_{xx} = G\chi,_{yy}, \quad \sigma_{xy} = -G\chi,_{xy}, \quad \sigma_{yy} = G\chi,_{xx}. \qquad (2.3)$$

The strain components then defined in terms of χ through (2.1) must satisfy the strain compatibility condition

$$\varepsilon_{xx,yy} + \varepsilon_{yy,xx} = 2\varepsilon_{xy,xy}. \qquad (2.4)$$

This means that χ must satisfy the generalized biharmonic equation

$$\left(\frac{\partial^2}{\partial x^2} + \varepsilon^2 \frac{\partial^2}{\partial y^2}\right)\left(c^2 \frac{\partial^2}{\partial x^2} + \frac{\partial^2}{\partial y^2}\right)\chi = 0. \tag{2.5}$$

Here $1/\varepsilon^2$ and c^2 are the larger and smaller roots λ of the equation

$$A\lambda^2 - (1-2B)\lambda + C = 0, \tag{2.6}$$

and

$$A = G/E, \quad B = \nu G/E, \quad C = G/E'. \tag{2.7}$$

We wish to consider fibrous materials with the fibers aligned parallel to the x-direction. For such materials the tensile modulus E in the fiber direction is much larger than the shear modulus G, so the coefficients A and B in (2.6) are small. In such cases the roots of (2.6) are approximately 1/A and C, so we have

$$\varepsilon^2 \cong G/E \quad \text{and} \quad c^2 \cong G/E'. \tag{2.8}$$

Then ε is a small parameter and c is of normal order. With an abuse of language, we call ε the *fiber extensibility* and c the *normal compressibility*.

In the case of fabricated materials with fibers running in both the x and y directions, both ε and c are small parameters. The approximation methods that can be used in such cases are much simpler than those available when there is only one small parameter, but for our present purpose, which is to examine the nature of these methods as asymptotic approximations to elasticity theory, it is clearer to restrict attention to cases in which ε is a small parameter but c is not.

As it stands, (2.5) is elliptic so its solutions are all absolutely smooth. If we set $\varepsilon = 0$ directly in the equation, we obtain

$$(c^2\partial^2/\partial x^2 + \partial^2/\partial y^2)\partial^2\chi/\partial x^2 = 0, \tag{2.9}$$

a parabolic equation with real characteristics along the lines y = constant, the fibers. To carry matters to an extreme, when c is also small the equation obtained by setting c = 0 is

$$\partial^4\chi/\partial x^2\partial y^2 = 0, \tag{2.10}$$

which is hyperbolic, with both the x and y directions as characteristics. The solution of the latter equation,

$$\chi = yf_1(x) + xf_2(y) + f_3(x) + f_4(y), \tag{2.11}$$

can have discontinuities in derivatives across lines x = constant or y = constant. As we shall see, solutions of the parabolic equation (2.9) can also have discontinuities across fibers.

The discontinuities exhibited by solutions of (2.9) provide important information about solutions of the exact equation (2.5). Just because solutions of the exact equation are infinitely differentiable, regions in which stresses are very large or changing very rapidly often do not stand out clearly even when exact analytical solutions can be obtained. The theory based on the assumption of fiber inextensibility brings the most important features of solutions into sharp focus and, since it is a simpler theory, makes it possible to solve problems that might otherwise be too difficult to attack analytically.

3. Solution in complex variables [3].

In order to understand the meaning of the singularities that arise in solutions based on the inextensible (ε=0) theory, we shall examine some exact elastic solutions in the limit as ε approaches zero. Provided that $\varepsilon \neq 1/c$, the general solution of (2.5) has the form

$$\chi = \chi_1(x,cy) + \chi_2(\varepsilon x,y), \tag{3.1}$$

where χ_1 is a harmonic function of x and cy, and χ_2 is harmonic in εx and y. It is convenient to introduce the complex variables

$$z_1 = x + icy \quad\text{and}\quad z_2 = \varepsilon x + iy \tag{3.2}$$

and write

$$\chi_1 = \text{Re}\ \phi(z_1) \quad \text{and} \quad \chi_2 = \text{Re}\ \psi(z_2), \tag{3.3}$$

where Re means the real part, and ϕ and ψ are analytic. Then the expressions for the stress components (2.3) take the forms

$$\begin{aligned} \sigma_{xx}/G &= \text{Re}(-c^2\phi'' - \psi''), \\ \sigma_{xy}/G &= \text{Re}(-ic\phi'' - i\varepsilon\psi''), \\ \sigma_{yy}/G &= \text{Re}(\phi'' + \varepsilon^2\psi''). \end{aligned} \tag{3.4}$$

The displacement components u and v are to be obtained from (2.1). Expressing the strains in terms of displacements, and at the same time expressing the elastic constants in terms of the fundamental parameters ε and c, we write (2.1) as

$$\begin{aligned} Gu_{,x} &= a\varepsilon^2\sigma_{xx} - b\varepsilon^2\sigma_{yy}, \\ Gv_{,y} &= ac^2\sigma_{yy} - b\varepsilon^2\sigma_{xx}, \\ G(u_{,y}+v_{,x}) &= \sigma_{xy}. \end{aligned} \tag{3.5}$$

Recalling the approximate expressions (2.8) for ε and c, we see that $a \cong 1$ and $b \cong \nu$. The four coefficients a, b, c and ε are connected by the relation

$$a = (1-2b\varepsilon^2)/(1+\varepsilon^2c^2). \tag{3.6}$$

By using (3.4) in (3.5), integrating (with the use of (3.6)), and absorbing the integration constants into ϕ and ψ, we obtain

$$\begin{aligned} u &= -(\partial/\partial x)\text{Re}[\psi+k\varepsilon^2(\phi-\psi)], \\ v &= -(\partial/\partial y)\text{Re}[\phi+k\varepsilon^2(\psi-\phi)]. \end{aligned} \tag{3.7}$$

Here the parameter k is defined by

$$k = b+ac^2. \tag{3.8}$$

4. The inextensible theory [4,5].

The *inextensible* theory is the theory based on setting $\varepsilon = 0$ in the equations while treating all unknowns and their derivatives as independent of ε. This amounts to the same thing as assuming that all unknowns can be expanded in powers of ε and retaining only the first term of each series. In the jargon of asymptotic analysis, this is an "outer" approximation.

Setting $\varepsilon = 0$ in the elastic solution (3,4), (3.7), is a little touchy since the parametric dependence of the potentials ϕ and ψ on ε is not known. It is clearer to return to the stress-strain relations (3.5) and the equilibrium equations

$$\sigma_{xx,x} + \sigma_{xy,y} = 0, \quad \sigma_{yx,x} + \sigma_{yy,y} = 0. \tag{4.1}$$

By setting $\varepsilon = 0$ in (3.5) we obtain

$$u,_x = 0 \quad \text{and} \quad Gv,_y = c^2\sigma_{yy}, \tag{4.2}$$

with the relation between shearing stress and shearing strain unaltered. The first of these equations implies that $u = u(y)$; each fiber y moves with no stretching. The shearing stress then takes the form

$$\sigma_{xy}/G = u'(y) + v,_x(x,y). \tag{4.3}$$

Substitution of (4.2b) and (4.3) into (4.1b) then yields

$$c^2v,_{xx} + v,_{yy} = 0. \tag{4.4}$$

Thus v is a harmonic function of x and cy. This agrees with the result obtained by setting $\varepsilon = 0$ in (3.7b).

Let us consider problems in which v or σ_{yy} is specified at each boundary point, so that we have a standard Dirichlet, Neumann, or mixed problem for v. Let us regard v as having been determined.

The tensile stress σ_{xx} no longer appears in the reduced stress-strain relations (4.2) because it is an arbitrary reaction to the constraint of fiber inextensibility. Indeed, as we shall see, σ_{xx} can be infinite because this does not imply infinite strain or displacement. The reaction σ_{xx} is determined from the equilibrium equation (4.1a), which yields

$$\sigma_{xx} = -Gxu''(y) - Gv,_y(x,y) + f(y). \tag{4.5}$$

We expect to be able to impose boundary conditions of the same type that would lead to a well-set problem in elasticity theory, but one restriction is immediately obvious. The displacement component u cannot be required to have different values at two ends of the same fiber. From (3.5a), σ_{xx} would be $O(1/\varepsilon^2)$ in such problems, contrary to our assumption in treating σ_{xx} as $O(1)$.

To proceed with the specification of a problem in the simplest possible way, let us suppose that u is specified at one end of each fiber, so that u(y) is known directly from the boundary conditions. With v or σ_{yy} known at each boundary point, we complete the specification by giving σ_{xx} (or the x-component of traction) at every boundary point where u was not specified. This determines the function f(y) in (4.5) and completes the solution of the problem.

When a boundary lies along a fiber, so that data are specified along a characteristic, one should expect some kind of trouble. In a problem of the kind outlined, there is no difficulty connected with specifying v or σ_{yy} along a boundary fiber. The interesting feature of the solution comes from specifying the x-component of traction. Suppose that the boundary lies along the fiber y = 0, with the body above it, and the shearing stress is specified as $\sigma(x)$ along the boundary. From (4.3),

$$\sigma_{xy}(x,0+) = u'(0) + v,_x(x,0), \tag{4.6}$$

and since u and v have already been determined without use of the shearing stress boundary condition, that condition apparently cannot be satisfied, although it can be satisfied in the exact elastic solution.

The solution to this difficulty is simply to recognize that the shearing stress is discontinuous across the fiber y = 0, so that in the vicinity of $y = 0$ the stress has the form

$$\sigma_{xy} = \sigma(x) + H(y)[\sigma_{xy}(x,0+)-\sigma(x)]+ \ldots, \tag{4.7}$$

where $H(y)$ is the unit step function. This does not affect any equation except the x-component of the equilibrium equation, which becomes

$$\sigma_{xx,x} = -\delta(y)[\sigma_{xy}(x,0+)-\sigma(x)]+\ldots, \tag{4.8}$$

where $\delta(y)$ is the Dirac delta. Thus the tensile stress becomes infinite in the boundary fiber, and has the form

$$\sigma_{xx} = F(x)\delta(y)+ \ldots, \tag{4.9}$$

where $F(x)$ represents the finite force carried by the boundary fiber. It is related to the shearing stress discontinuity by

$$F'(x) = \sigma(x)-\sigma_{xy}(x,0+). \tag{4.10}$$

We call such fibers *singular fibers*.

5. Outer and inner limits [2].

Let us examine a solution with a singular fiber and compare it to the exact elastic solution. We consider a half-space $y \geqq 0$ with a sinusoidally varying tangential load on the boundary $y = 0$, and no stress or displacement at infinity:

$$\sigma_{xy}(x,0) = \sigma \sin(x/L), \quad \sigma_{yy}(x,0) = 0. \tag{5.1}$$

The second of these conditions, with (4.2b), is $v,_y = 0$, and with $v = 0$ at infinity it then follows that v, being harmonic, is zero. In the inextensible solution $u(y)$ is identically zero since it vanishes at $x = \pm\infty$. It follows easily that all stress components are zero, except σ_{xx} in the boundary fiber $y = 0$. With $\sigma_{xy}(x,0+) = 0$, the force in the boundary fiber is found from (4.10) and (5.1a) to be

$$F(x) = -\sigma L \cos(x/L), \tag{5.2}$$

plus an arbitrary constant.

In the exact elastic theory the conditions of the problem are satisfied by the potentials

$$\phi = \varepsilon D e^{iz_1/L} \quad \text{and} \quad \psi = -\varepsilon D e^{iz_2/\varepsilon L}, \tag{5.3}$$

where

$$D = \sigma L^2/G(1-\varepsilon c). \tag{5.4}$$

Then

$$(\sigma_{xx}/\sigma)(1-\varepsilon c) = [-(1/\varepsilon)e^{-y/\varepsilon L}+\varepsilon c^2 e^{-cy/L}]\cos(x/L), \tag{5.5}$$

$$(\sigma_{xy}/\sigma)(1-\varepsilon c) = [e^{-y/\varepsilon L}-\varepsilon c e^{-cy/L}]\sin(x/L), \tag{5.6}$$

$$(\sigma_{yy}/\sigma)(1-\varepsilon c) = [\varepsilon e^{-y/\varepsilon L}-\varepsilon e^{-cy/L}]\cos(x/L), \tag{5.7}$$

$$(u/L)(G/\sigma)(1-\varepsilon c) = \varepsilon[(1-k\varepsilon^2)e^{-y/\varepsilon L}-k\varepsilon^2 e^{-cy/L}]\sin(x/L), \tag{5.8}$$

$$(v/L)(G/\sigma)(1-\varepsilon c) = \varepsilon[-k\varepsilon e^{-y/\varepsilon L}+c(1-k\varepsilon^2)e^{-cy/L}]\cos(x/L). \tag{5.9}$$

The *outer* limit of the elastic solution is obtained by letting ε approach zero with y fixed. In this limit we obtain $u = v = \sigma_{yy} = 0$ for all y. The limit of σ_{xy} is discontinuous:

$$\sigma_{xy}(x,0) = \sigma \sin(x,L), \quad \sigma_{xy}(x,y) = 0 \quad (y > 0). \tag{5.10}$$

The limit of σ_{xx} is zero for all $y > 0$, but at $y = 0$ its limit is infinite. However, the sequence of functions

$$e^{-y/\varepsilon L}(\varepsilon L)^{-1} \tag{5.11}$$

defines a Dirac delta, so we may write the limit of σ_{xx} as $F(x)\delta(y)$, where F(x) is given by (5.2). Thus, all of the strange features of the solution according to the inextensible theory are found to be correct representations of the outer limit of the elastic solution.

The outer limit of the elastic solution was obtained with almost no computation by using the inextensible theory, but this limit does not give the details of the solution in the stress concentration layer near the boundary. To examine this layer, we introduce the stretched coordinate

$$Y = y/\varepsilon \tag{5.12}$$

and the scaled variables defined by

$$\sigma_{xx} = (1/\varepsilon)\bar{\sigma}_{xx}, \quad \sigma_{xy} = \bar{\sigma}_{xy}, \quad u = \varepsilon\bar{u}. \tag{5.13}$$

The *inner* limit of the elastic solution is found by letting ε approach zero with Y fixed. In this limit we obtain

$$\bar{\sigma}_{xx}/\sigma = -e^{-Y/L}\cos(x/L), \tag{5.14}$$

$$\bar{\sigma}_{xy}/\sigma = e^{-Y/L}\sin(x/L), \tag{5.15}$$

$$\bar{u}/L = (\sigma/G)e^{-Y/L}\sin(x/L), \tag{5.16}$$

and $v = \sigma_{yy} = 0$.

In this limit the shearing stress is continuous, bridging the gap between the values given by the discontinuous outside limit.

The limiting values of σ_{yy} and v are independent of Y, and equal to the boundary values of σ_{yy} and v given by the outer limit. This is trivial in the present case, but stated in this way, it is a general feature of stress concentration layers.

The rapid variation of shearing stress across the boundary layer requires for its equilibration a very high tensile stress, of order $1/\varepsilon$. This is produced by the small but non-zero stretching $\varepsilon\bar{u}_{,x}$.

6. Boundary stress concentration layers [6].

Let us consider the general case in which part of the boundary of a body lies along a fiber. As in the preceding example, let the bounding fiber be on $y = 0$, with the body in the region $y \geqq 0$. Suppose that the solution from the inextensible theory is known, and mark quantities from that solution (for $y > 0$) with a superscript "o".

For stress concentration layers laying along boundaries it is convenient to express all unknowns as modifications added to the values given by the inextensible solution. Using the example in Section 5 as a guide, we write

$$\sigma_{xx} = \sigma^{o}_{xx}(x,y) + (1/\varepsilon)\bar{\sigma}_{xx}(x,Y), \tag{6.1}$$

$$\sigma_{xy} = \sigma^{o}_{xy}(x,y) + \bar{\sigma}_{xy}(x,Y), \tag{6.2}$$

$$\sigma_{yy} = \sigma^{o}_{yy}(x,y) + \varepsilon\bar{\sigma}_{yy}(x,Y), \tag{6.3}$$

$$u = u^{o}(y) + \varepsilon\bar{u}(x,Y), \tag{6.4}$$

$$v = v^{o}(x,y) + \varepsilon\bar{v}(x,Y). \tag{6.5}$$

In principle both the outer quantities (f^{o}, say) and the boundary corrections ($\bar{f}$) should be regarded as depending parametrically on ε, and we suppose that they can be expanded in powers of ε. The equations that we intend to obtain actually refer to the first term of the series expansion in each case.

The distinguishing feature of the boundary corrections $\bar{f}$ is that they are regarded as functions of $Y = y/\varepsilon$, so that $\bar{f}_{,y} = \bar{f}_{,Y}/\varepsilon$; that is, we suppose that the y-derivative of a boundary correction function is $O(1/\varepsilon)$ in comparison to the function itself.

On substituting (6.1) to (6.3) into the equilibrium equations and using the fact that the outer quantities satisfy those equations by themselves, we find that

$$\bar{\sigma}_{xx,x} + \bar{\sigma}_{xy,Y} = 0 \quad \text{and} \quad \bar{\sigma}_{yx,x} + \bar{\sigma}_{yy,Y} = 0. \tag{6.6}$$

By using (6.1) to (6.5) in the stress-strain relations (3.5) and taking into account the relations (4.2) satisfied by the inextensible solution, on taking the inner limit we obtain

$$\bar{\sigma}_{xx} = G\bar{u}_{,x} \quad \text{and} \quad \bar{\sigma}_{xy} = G\bar{u}_{,Y}. \tag{6.7}$$

In addition, (3.5b) yields $\bar{v}_{,Y} = 0$. This means that v has no variation through the thickness of the boundary layer, even to $O(\varepsilon)$, except that included in the outer approximation. The term $\bar{\sigma}_{yy}$ in (6.3) is also insignificant. We could determine it by using (6.6b), but since it is of no importance we merely ignore (6.6b).

The important equilibrium equation is (6.6a). By using (6.7) in it we find that $\bar{u}$ is a harmonic function of x and Y:

$$\bar{u}_{,xx} + \bar{u}_{,YY} = 0. \tag{6.8}$$

In the exact elastic solution (3.7a), terms of this kind are contributed by the potential $\psi(\varepsilon x+iy)$.

Now consider the boundary conditions. Let $\sigma(x)$ be the prescribed shearing stress on the boundary $y = 0$. By using (6.2) we obtain

$$\bar{\sigma}_{xy}(x,0) = \sigma(x) - \sigma^{0}_{xy}(x,0+). \tag{6.9}$$

Taking the outer limit in (6.2) and using the definition of σ^{0}_{xy} as the outer limit of σ_{xy}, and the fact that the outer limit of $\bar{\sigma}_{xy}(x,y/\varepsilon)$ is $\bar{\sigma}_{xy}(x,\infty)$, we obtain

$$\bar{\sigma}_{xy}(x,\infty) = 0. \tag{6.10}$$

If the boundary fiber runs from $x = 0$ to $x = L$, say, then conditions at these values of x are also needed. Suppose that u is specified along a boundary $x = x(y)$ that intersects the boundary fiber at $x(0) = x_0$. Then $\bar{u}$ must vanish on that boundary since $u^{0}(y)$ already has the prescribed values. In terms of x and Y, the boundary is at $x = x(\varepsilon Y)$, so in the limit as ε approaches zero with Y fixed, the boundary is at $x = x(0)$ for all Y. Thus,

$$\bar{u}(x_0,Y) = 0 \quad (0 \leqq Y < \infty). \tag{6.11}$$

At an end where finite traction is specified, $\bar{\sigma}_{xx}$ must vanish since otherwise σ_{xx} would become infinite as ε approaches zero:

$$\bar{\sigma}_{xx}(x_0,Y) = 0 \quad (0 \leqq Y < \infty). \tag{6.12}$$

This corresponds to the condition in the inextensible theory that the force F(x) in a singular fiber must vanish at an end where the prescribed traction is finite.

Let us define the total tensile force in the boundary layer as

$$\bar{F}(x) = \int_0^\infty \bar{\sigma}_{xx} dY. \tag{6.13}$$

By integrating (6.6a) with respect to Y and using (6.9), (6.10), and (4.10), we obtain

$$\bar{F}'(x) = \sigma(x) - \sigma^0_{xy}(x,0+) = F'(x), \tag{6.14}$$

where F(x) is the total tensile force given by the inextensible theory. If the fiber has an end where finite traction is prescribed, both F and $\bar{\sigma}_{xx}$ vanish there, so $F = \bar{F}$. If the fiber has no end, as in the example in Section 5, F is indeterminate to the extent of an arbitrary constant, so for some choice of the constant we have $F = \bar{F}$.

In pure traction boundary value problems, with finite tractions at x = 0 and x = L, say, F must vanish at both of these points. This is important for the existence of solutions of (6.8). For, with boundary conditions

$$\bar{u}_{,Y}(x,0) = F'(x)/G, \quad \bar{u}_{,Y}(x,\infty) = \bar{u}_{,x}(0,Y) = \bar{u}_{,x}(L,Y) = 0, \tag{6.15}$$

the problem is a Neumann problem and no solution exists unless the integral of the normal derivative around the boundary is zero. But this existence condition merely requires that F(0) = F(L), so it is satisfied if F vanishes at both ends.

7. Example: Sheared sheet [6].

In the example discussed in Section 1, considered as a plane stress problem, the inextensible theory yields

$$u^0 = 0, \quad v^0 = (\sigma/G)x, \quad \sigma^0_{yy} = 0, \quad \sigma^0_{xy} = \sigma, \tag{7.1}$$

and

$$\sigma^0_{xx} = F(x)[\delta(y)-\delta(y-H)], \quad F(x) = \sigma(L-x). \tag{7.2}$$

Let us use the theory of stress concentration layers to examine the singular fiber along y = 0.

The boundary conditions on $\bar{u}$ are

$$\bar{u}_{,Y}(X,0) = -\sigma/G,\ \bar{u}_{,Y}(X,\infty) = \bar{u}(0,Y) = \bar{u}_{,X}(L,Y) = 0. \tag{7.3}$$

The solution is easily found by separation of variables, and the series defining the stress components can be summed to give

$$\bar{\sigma}_{xx} - i\bar{\sigma}_{xy} = i\sigma - (2\sigma/\pi)\ln[\tan(\pi Z/4L)], \tag{7.4}$$

where

$$Z = X + iY = z_2/\varepsilon. \tag{7.5}$$

The boundary layer solution shows that the stress has a weak singularity at $Z = 0$, which was concealed inside the singular fiber in the inextensible solution:

$$\bar{\sigma}_{xx} \sim (2\sigma/\pi)\ln(L/R),\quad R^2 = X^2 + Y^2. \tag{7.6}$$

A singularity of the same form occurs whenever a singular fiber intersects a boundary on which u is prescribed, with σ replaced by the value of $-F'$ there.

8. Interior stress concentration [2].

Stress concentration layers can also occur in the interior of a body, for a variety of reasons. A point force may be applied to the boundary, a fiber may run partly along the boundary and partly interior to the body, or a fiber may have a point in common with a reentrant part of the boundary, such as the tip of a crack. In the latter cases the singular fiber predicted by the inextensible theory is a line across which $u'(y)$ and σ_{xy} are discontinuous. From (4.5) we see that a jump discontinuity in $u'(y)$ produces a Dirac delta in σ_{xx}.

For a simple example, consider a half-space $x \geqq 0$ loaded by a point force normal to the boundary,

$$\sigma_{xx}(0,y) = F\delta(y),\quad \sigma_{xy}(0,y) = 0, \tag{8.1}$$

and no displacement at infinity. Within the inextensible theory the solution is $u = v = 0$ everywhere, $\sigma_{xy} = \sigma_{yy} = 0$ everywhere, and $\sigma_{xx} = F\delta(y)$ everywhere. The boundary load penetrates along the fiber $y = 0$ with no attenuation.

The exact elastic solution of the same problem (with zero stress rather than zero displacement at infinity) is furnished by the stress potentials

$$\phi' = \varepsilon D \ln(z_1/L), \quad \psi' = -D \ln(z_2/\varepsilon L) \tag{8.2}$$

in the notation of Section 3. Here

$$D = F/G\pi(1-\varepsilon c). \tag{8.3}$$

Then

$$\sigma_{xx}/GD = \frac{\varepsilon x}{(\varepsilon x)^2+y^2} - \frac{c^2 \varepsilon x}{x^2+(cy)^2}, \tag{8.4}$$

$$\sigma_{xy}/GD = \frac{\varepsilon y}{(\varepsilon x)^2+y^2} - \frac{c^2 \varepsilon y}{x^2+(cy)^2}, \tag{8.5}$$

$$\sigma_{yy}/GD = \frac{\varepsilon x}{x^2+(cy)^2} - \frac{\varepsilon^3 x}{(\varepsilon x)^2+y^2}, \tag{8.6}$$

$$u/D = \varepsilon(1-k\varepsilon^2)\ln(r_2/\varepsilon L) - k\varepsilon^3 \ln(r_1/L) \tag{8.7}$$

$$v/D = \varepsilon c(1-k\varepsilon^2)\theta_1 - k\varepsilon^2\theta_2, \tag{8.8}$$

where

$$r_1 e^{i\theta_1} = x + icy \quad \text{and} \quad r_2 e^{i\theta_2} = \varepsilon x + iy. \tag{8.9}$$

In the outer limit of the elastic solution, obtained by letting ε approach zero with y fixed, all variables vanish except σ_{xx}, and σ_{xx} vanishes everywhere except on $y = 0$. On that line σ_{xx} diverges to infinity. The singular behavior of σ_{xx} is in the term

$$\frac{1}{\pi}\frac{\varepsilon x}{(\varepsilon x)^2+y^2}, \tag{8.10}$$

whose integral with respect to y is unity. Thus, the limit of σ_{xx} can be written as $F\delta(y)$.

In the present problem there is no characteristic length; the length L in (8.2) is an arbitrary length supplied to make the arguments of the logarithms be numbers. Then at the place x along the singular fiber there is no characteristic length except x itself. For this reason we should expect the thickness of the stress concentration layer to be of the order of εx and the tensile stress in it to be of the order of $F/\varepsilon x$. The latter estimate is easily verified. From (8.4), the tensile stress at $y = 0$ is $F/\pi\varepsilon x$ to leading order in ε.

The inner limit is obtained by introducing the stretched coordinate $Y = y/\varepsilon$ and the scaled variables (5.13). Then letting ε approach zero with Y fixed, we obtain

$$\bar{\sigma}_{xx} = (F/\pi)x/R^2, \; \bar{\sigma}_{xy} = (F/\pi)Y/R^2, \; \bar{u} = (F/\pi G)\ell n(R/L), \tag{8.11}$$

where

$$R^2 = x^2 + Y^2. \tag{8.12}$$

The tensile stress $\bar{\sigma}_{xx}$ is constant on the circles $(x-x_0)^2 + Y^2 = x_0^2$, which in terms of x and y are ellipses with semi-axes x_0 and εx_0. Thus if we define the stress concentration layer to be the region in which $\bar{\sigma}_{xx}$ exceeds some particular value, the thickness of the layer at $x=x_0$ is $2\varepsilon x_0$, and it is in this sense that the thickness of the layer grows in proportion to x. Alternatively, the half-width of the distribution of $\bar{\sigma}_{xx}$ as a function of Y can be defined as the value of Y at the inflection point, $Y = Y_1 x$, where Y_1 is the value at $x = 1$.

9. Interior stress concentrations. General. [3,7]

Let us consider the general case of a stress concentration layer interior to a body. Let the singular fiber given by the inextensible theory lie along $y = 0$. The tensile force in the singular fiber is connected to the discontinuity in $u'(y)$ across it by

$$F'(x) = -G[u'(0+)-u'(0-)] \equiv -G\Delta u', \tag{9.1}$$

so

$$F(x) = F_0-(G\Delta u')x. \tag{9.2}$$

In the inner limit v and σ_{yy} approach the same values given by the outer limit. For u we write

$$u = u(0) + \varepsilon\bar{u}(x,Y), \tag{9.3}$$

where u(0) is the value given by the inextensible solution. The shearing stress is

$$\sigma_{xy} = \bar{\sigma}_{xy}(x,Y) + Gv,_x(x,y), \tag{9.4}$$

where

$$\bar{\sigma}_{xy} = G\bar{u},_Y. \tag{9.5}$$

As usual, we write

$$\sigma_{xx} = (1/\varepsilon)\bar{\sigma}_{xx}. \tag{9.6}$$

The x-component of the equilibrium equation reduces to (6.8) and implies that $\bar{u}$ is a harmonic function of x and Y.

Since $\bar{u}$ here is not an additive correction to the inextensible solution u(y), but rather a smooth function rounding out the discontinuity in u'(y), the boundary conditions on $\bar{u}$ at $Y = \pm\infty$ are

$$\bar{u},_Y(x,\pm\infty) = u'(0\pm). \tag{9.7}$$

These conditions are obtained most easily by assuming that there are matching regions in which σ_{xy} is given in the limit both by (9.4) and by the outer approximation. Then setting $Y = \pm L/\sqrt{\varepsilon}$ and $y = \pm\sqrt{\varepsilon}L$ yields (9.7) in the limit.

As in the case of edge layers, the boundary condition at an end $x = x_0$ where finite traction is prescribed is that $\bar{\sigma}_{xx}$ is zero there. In terms of $\bar{u}$,

$$\bar{u},_x(x_0,Y) = 0, \quad -\infty < Y < \infty. \tag{9.8}$$

At an end where u is specified,

$$\bar{u}(x_0,Y) = u'(0+)Y \ (Y \geqq 0), \ u'(0-)Y \ (Y \leqq 0). \tag{9.9}$$

10. Interior layer through crack tip [7,8].

Consider a finite body with a traction-free crack running from the boundary to some interior point. Let the origin of coordinates be at the crack tip and suppose that the crack lies along a curve $x = x_c(y)$ in the third quadrant $(x < 0, y < 0)$. For the displacement $u(y)$ given by the inextensible theory we use the notation

$$u_0(y) \ (y \geqq 0),\ u_+(y) \ (y \leqq 0,\ x \geqq x_c),$$
$$u_-(y) \ (y \leqq 0,\ x \leqq x_c). \tag{10.1}$$

There is generally a singular fiber along the line $y = 0$ through the crack tip. The force in it satisfies

$$F'(x) = -G[u_0'(0) - u_\pm'(0)] \equiv -G\Delta u_\pm' . \tag{10.2}$$

If the fiber $y = 0$ intersects the outer boundary at $x = -L_-$ and $x = L_+$, and the applied traction is finite at these points, so that F must vanish there, we obtain

$$F(x) = \begin{cases} -G\Delta u_+'(x-L_+) \ (x \geqq 0), \\ -G\Delta u_-'(x+L_-) \ (x \leqq 0). \end{cases} \tag{10.3}$$

The force F_0 at the crack tip is

$$F_0 = -GL_-\Delta u_-' = GL_+\Delta u_+' . \tag{10.4}$$

Continuity of F at the tip, which gives the two alternative expressions for F_0 in (10.4), is easily seen to be equivalent to the equilibrium condition that the resultant shearing forces on the lines $y = 0+$ and $y = 0-$ are equal. If there is no crack, $u_+ = u_-$ and (10.4) implies that $\Delta u' = 0$, so $F = 0$.

Although the fiber force is largest at the crack tip, the inextensible theory conceals the infinite stress singularity that can be expected at the tip. In the stress concentration layer represented by the singular fiber, $\bar{u}$ is harmonic and satisfies the boundary conditions

$$\bar{u}_{,Y}(x,-\infty) = u_{\pm}'(0), \ \bar{u}_{,Y}(x,\infty) = u_0'(0), \tag{10.5}$$

$$\bar{u}_{,x}(\pm L_{\pm},Y) = 0 \ (\text{all } Y), \ \bar{u}_{,x}(0\pm,Y) = 0 \quad (Y \leqq 0). \tag{10.6}$$

In (10.5a), ± denotes the sign of x. In (10.6b) we have assumed that the crack is not tangential to the x-axis at its tip, for if it were, the crack boundary would not appear to be at x = 0 for all Y < 0.

Since $\bar{u}$ is harmonic in x and Y we can write

$$\bar{u} = u_0'(0)Y + \text{Re } U(Z), \ Z = x + iY, \tag{10.7}$$

where U(Z) is analytic. Then the boundary conditions become

$$\text{Re}[iU'(x-i\infty)] = -\Delta u_{\pm}', \quad \text{Re}[iU'(x+i\infty)] = 0, \tag{10.8}$$

and

$$\text{Re}[U'(\pm L_{\pm}+iY)] = 0, \quad \text{Re}[U'(\pm 0+iY)] = 0 \quad (Y \leqq 0). \tag{10.9}$$

From (10.8b) and the Cauchy-Riemann equations, U' approaches a constant value in the region at x+i∞, and from (10.8b) and (10.9a) together, this value is zero:

$$U'(x+i\infty) = 0. \tag{10.10}$$

It follows by a similar argument from (10.8a) and (10.9) that

$$U'(x-i\infty) = i\Delta u_+'(x > 0), \ U'(x-i\infty) = i\Delta u_-' \ (x < 0). \tag{10.11}$$

The intensity of the stress singularity at the tip of the crack is the quantity of primary interest in the problem, and it can be evaluated directly from the boundary conditions by using a J-integral technique. Near the crack tip, U(Z) has the form

$$U(Z) = 2A(iZ)^{1/2}, \tag{10.12}$$

to lowest order in Z, where A is a real constant. To verify that this satisfies the zero-traction conditions on the crack surfaces, we compute

$$\bar{\sigma}_{xx}/G = \text{Re}[A(i/Z)^{1/2}] = AR^{-1/2}\cos \tfrac{1}{2}(\theta - \tfrac{\pi}{2}), \tag{10.13}$$

where

$$Re^{i\theta} = x + iY. \tag{10.14}$$

The factor A measures the intensity of the stress singularity. We wish to determine A.

The integral $\int U'^2 dZ$ is path-independent since U' is analytic. If we use a path that is a circle of small radius around the crack tip, with (10.12) we obtain

$$\oint U'^2 dZ = \oint (A^2 i/Z)dZ = -2\pi A^2. \tag{10.15}$$

The integral can also be evaluated by first deforming the contour into a path that follows the boundaries except in the regions at $Y = \pm\infty$. The path goes down the right side of the crack to $Y = -\infty$, crosses to the boundary $x = L_+$, returns on that boundary to $Y = +\infty$, crosses to $x = -L_-$, goes back to $Y = -\infty$ on that boundary, crosses back to the left-hand side of the crack, and returns to the crack tip on that side. Now, according to (10.9) U'^2 is real on the vertical parts of the path, and dZ is pure imaginary on those parts, so the contribution of the vertical parts of the path is pure imaginary. The integrals over the horizontal parts of the path can be evaluated by using (10.10) and (10.11), and we see that they are real. Since the total integral (10.15) is real, it follows that the sum of the pure imaginary integrals over the vertical parts must be zero, and thus

$$\oint U'^2 dZ = -L_+(\Delta u_+')^2 - L_-(\Delta u_-')^2. \tag{10.16}$$

By using (10.4) and writing L* for the harmonic mean of L_+ and L_-, we can write (10.16) as

$$\oint U'^2 dZ = -(2/L^*)(F_o/G)^2. \tag{10.17}$$

Then comparison with (10.15) gives

$$A = F_o/G(\pi L^*)^{1/2}. \tag{10.18}$$

ACKNOWLEDGMENT

This paper was prepared with the support of a grant MCS 79-03392 from the National Science Foundation. We gratefully acknowledge this support.

REFERENCES

[1] ROGERS, T.G. and PIPKIN, A.C. Small deflections of fiber-reinforced beams or slabs. J. Appl. Mech. 38, (1971) 1047-8.

[2] EVERSTINE, G.C. and PIPKIN, A.C. Stress channelling in transversely isotropic elastic composites. ZAMP 22, (1971) 825-34.

[3] PIPKIN, A.C. Stress analysis for fiber-reinforced materials. Advances in Appl. Mech. 19, (1979) 1-51.

[4] ENGLAND, A.H., FERRIER, J.E. and THOMAS, J.N. Plane strain and generalized plane stress problems for fibre-reinforced materials. J. Mech. Phys. Solids 21, (1973) 279-301.

[5] MORLAND, L.W. A plane theory of inextensible transversely isotropic elastic composites. Int. J. Solids Struct. 9, (1973) 1501-18.

[6] EVERSTINE, G.C. and PIPKIN, A.C. Boundary layers in fiber-reinforced materials. J. Appl. Mech. 40, (1973) 518-22.

[7] SPENCER, A.J.M. Boundary layers in highly anisotropic plane elasticity. Int. J. Solids Struct. 10, (1974) 1103-23.

[8] SANCHEZ, V. and PIPKIN, A.C. Crack-tip analysis for elastic materials reinforced with strong fibers. Q. J. Mech. Appl. Math. 31, (1978) 349-62.

V
PROBLEMS FOR HELICALLY WOUND CYLINDERS

T. G. ROGERS
Department of Theoretical Mechanics
University of Nottingham
Nottingham, NG7 2RD
England

1. BALANCED ANGLE-PLY CYLINDERS

1.1 Introduction

This portion of the book deals with boundary value problems involving the deformation of hollow cylinders which are continuously reinforced throughout by two families of helically wound fibres. This type of cross-ply reinforcement can be realised fairly easily in practice by winding on a drum consecutively in opposing directions, and is sometimes termed *balanced angle-ply*. It has practical applications in the construction of pressure vessels, pipes, shafts and axles.

Until recently most of the relevant stress analysis in the literature has considered the reinforcement to lie in discrete cylindrical layers; alternatively the fibre-reinforced material has been modelled as an orthotropic elastic body [1] usually leading to complicated analysis for even the simplest configurations. The need to develop a continuum theory which is sufficiently simple to deal with more involved cylindrical components and is yet able to give reasonable qualitative predictions for such highly anisotropic behaviour has led to the *ideal fibre-reinforced material (IFRM)* being used in this context. This chapter gives an outline of the theory and the progress made so far.

In most practical applications the reinforcement is intended to inhibit large flexure and displacements of the components. Furthermore, the analysis is much easier if we assume small deformations. Consequently most of the theory we shall describe will deal with linear elastic behaviour; the remainder will concern small plastic-elastic deformations and, briefly, the relevant theory of large deformations.

The actual configuration considered (see Figure 1) is a hollow circular cylinder whose reinforcement at every point is directed at an angle $\pm\phi$ with the axis of the cylinder. The pitch ϕ of the helices sometimes depends on the radial coordinate r, but seldom on the other cylindrical polar coordinates θ and z. For convenience we shall assume that ϕ is constant throughout the cylinder, except when stated otherwise.

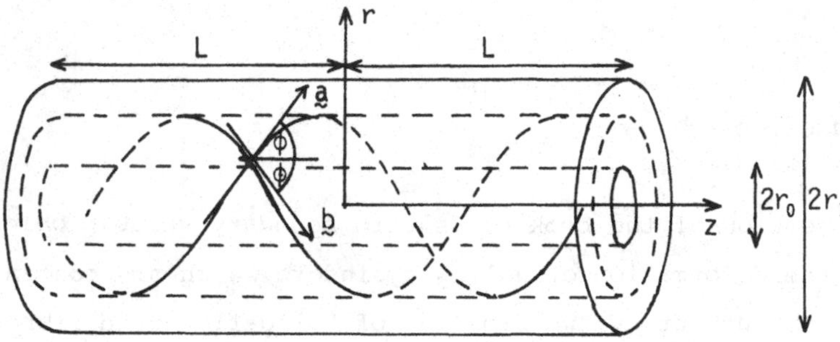

Figure 1. Configuration of cylinder, showing typical $\underset{\sim}{a}$ and $\underset{\sim}{b}$ fibres

We consider (again for convenience) only boundary tractions which are normal to the boundaries and symmetrical with respect to the normal cross-section z = 0. Thus for a cylinder of finite length 2L and inner and outer radii r_0 and r_1, the boundary conditions are specified as

$$\sigma_{rr} = \begin{cases} -p_0(z) & \text{on } r = r_0, \\ -p_1(z) & \text{on } r = r_1, \end{cases}$$

$$\sigma_{r\theta} = \sigma_{rz} = 0 \qquad \text{on } r = r_0, \quad r = r_1,$$

$$\sigma_{zz} = N(r,\theta) \qquad \text{on } z = \pm L, \tag{1}$$

$$\sigma_{rz} = \sigma_{\theta z} = 0 \qquad \text{on } z = \pm L.$$

Here σ_{rr}, $\sigma_{r\theta}$, etc. denote stress components referred to the cylindrical polar coordinate system, and $p_0(z) = p_0(-z)$, $p_1(z) = p_1(-z)$. The boundary conditions include the cases of inflation, axial extension and pure bending; yet again for convenience we do not include problems involving twist. Despite the simplifications assumed, the configuration is still a relatively complicated one for conventional linear elasticity, involving strong local orthotropy with the planes of elastic symmetry varying from point to point. Only the simplicity of the IFRM theory allows *analytical* solutions to be obtained.

Just as for problems in plane strain, this theory predicts the occurrence of singular stress layers both within and on the boundaries of the stressed body. These sheets of singular stress can occur in surfaces containing the fibres and in surfaces of normal cross-sections of the cylinder. The *normal* layers are analogous to those found in plane strain, and do not occur if the constraint of incompressibility is relaxed; unlike plane strain, these layers are not normal to the fibres themselves. Again the singular sheets can be interpreted as thin layers through which the shear stress changes rapidly and in which the in-plane direct stresses are very large.

1.2 Kinematics of infinitesimal deformations

The components of the displacement $\underset{\sim}{u}$ in cylindrical polars are denoted by (u,v,w). The orientations of the two *fibre-directions* at each point are denoted by unit vectors $\underset{\sim}{a}$ and $\underset{\sim}{b}$, whose components referred to the cylindrical polar coordinate system are

$$\underset{\sim}{a} = (0, \sin\phi, \cos\phi), \qquad \underset{\sim}{b} = (0, -\sin\phi, \cos\phi). \tag{2}$$

It is convenient to refer also to vector and tensor components with respect to local cartesian axes, and we use indicial notation to denote such components. Thus (a_1, a_2, a_3) denote the cartesian components of $\underset{\sim}{a}$ and have the values given in (2) if we choose the x_2- and x_3-directions to lie along the bisectors $\underset{\sim}{a}-\underset{\sim}{b}$ and $\underset{\sim}{a}+\underset{\sim}{b}$ respectively. Similarly, components of the infinitesimal strain $\underset{\sim}{e}$, which are denoted by e_{rr}, $e_{\theta\theta}$, etc. when referred to the cylindrical polar axes, are denoted by e_{ij} $(i,j = 1,2,3)$ with respect to cartesian axes.

The constraint of inextensibility in the $\underset{\sim}{a}$ and $\underset{\sim}{b}$ directions imply (as in equation (17) of Chapter I), for small deformations, that

$$a_i a_j e_{ij} = 0, \qquad b_i b_j e_{ij} = 0, \tag{3}$$

and the constraint of incompressibility requires

$$e_{kk} = 0; \tag{4}$$

here we adopt the usual summation convention for repeated indices.

For balanced angle-ply systems, equations (2) - (4) give

$$\begin{aligned} &e_{22}\sin^2\phi \pm 2e_{23}\sin\phi\cos\phi + e_{33}\cos^2\phi = 0, \\ &e_{11} + e_{22} + e_{33} = 0, \end{aligned} \tag{5}$$

so that a kinematically possible deformation needs to satisfy

$$e_{11} = e(\cot^2\phi - 1), \qquad e_{22} = -e\cot^2\phi, \qquad e_{33} = e, \qquad e_{23} = 0, \tag{6}$$

with no restriction on the strain components e_{12} and e_{13}. In terms of the displacement components (u,v,w) these give

$$\frac{\partial u}{\partial r} = (\cot^2\phi - 1)\frac{\partial w}{\partial z}, \qquad \frac{u}{r} + \frac{1}{r}\frac{\partial v}{\partial \theta} = -\cot^2\phi\frac{\partial w}{\partial z}, \qquad \frac{\partial v}{\partial z} + \frac{1}{r}\frac{\partial w}{\partial \theta} = 0. \tag{7}$$

Solutions of these equations are given by

$$u = \frac{r}{t^2}\frac{\partial^2\psi}{\partial z^2} - \frac{1}{r}\frac{\partial^2\psi}{\partial\theta^2}, \qquad v = \frac{1}{r}\frac{\partial\psi}{\partial\theta}, \qquad w = -\frac{\partial\psi}{\partial z}, \tag{8}$$

where ψ satisfies the equation

$$\frac{\partial^3\psi}{\partial r\partial z^2} - \frac{t^2}{r^2}\frac{\partial^3\psi}{\partial r\partial\theta^2} + \frac{2-t^2}{r}\frac{\partial^2\psi}{\partial z^2} + \frac{t^2}{r^3}\frac{\partial^2\psi}{\partial\theta^2} = 0, \tag{9}$$

and

$$t = \tan\phi. \tag{10}$$

These solutions are the most general ones available for IFRM cylinders, and are independent of the constitutive equations relating stress to deformation. Important special cases, such as axial symmetry or pure bending, lead to (9) being greatly simplified, with easy solutions.

1.3 Stress and equilibrium

The kinematically admissible solutions (8) are also statically admissible in the sense that they can be maintained in equilibrium by applying appropriate boundary tractions. This is because the three kinematical constraints of incompressibility and fibre-inextensibilities give rise to stress reactions in the form of hydrostatic pressure and fibre-tensions. These reactions are arbitrary functions of position, and must be determined from the equilibrium equations. Thus, as explained in Chapter I, the total stress $\underset{\sim}{\sigma}$ at each point of the cylinder can be decomposed into reaction stresses and an extra-stress $\underset{\sim}{s}$:

$$\sigma_{ij} = -p\delta_{ij} + T_a a_i a_j + T_b b_i b_j + s_{ij}\ , \tag{11}$$

where s_{ij} is related to the deformation through the constitutive equations describing different material behaviours. Without loss of generality (again refer Chapter I), we can also write

$$a_i a_j s_{ij} = 0, \qquad b_i b_j s_{ij} = 0, \qquad s_{kk} = 0, \tag{12}$$

so that now s_{ij} is unaltered by the superposition of an arbitrary hydrostatic pressure or an arbitrary tension imposed in either fibre-direction. Note, however, that in general T_a is *not* the direct stress component in the $\underset{\sim}{a}$-direction; for balanced angle-ply materials this is given by

$$t_a = a_i a_j \sigma_{ij} = -p + T_a + T_b \cos^2 2\phi, \tag{13}$$

and similarly for the $\underset{\sim}{b}$-direction.

For balanced angle-ply reinforcement, (2) and (11) give

$$\begin{aligned}
\sigma_{11} &= -p + s_{11}, \\
\sigma_{22} &= -p + (T_a + T_b)\sin^2\phi + s_{22}, \\
\sigma_{33} &= -p + (T_a + T_b)\cos^2\phi + s_{33}, \qquad (14) \\
\sigma_{23} &= \tfrac{1}{2}(T_a - T_b)\sin 2\phi + s_{23}, \\
\sigma_{12} &= s_{12}, \qquad \sigma_{13} = s_{13},
\end{aligned}$$

where the extra-stress components are related through (12) in the form

$$s_{11} = S(\cot^2\phi - 1), \qquad s_{22} = -S\cot^2\phi, \qquad s_{33} = S, \qquad s_{23} = 0, \qquad (15)$$

analogous to the strain relations (6). For later use, we eliminate p, T_a and T_b from (14) to obtain

$$S = \frac{\sigma_{33} - \sigma_{11} - (\sigma_{22} - \sigma_{11})\cot^2\phi}{2(1 - \cot^2\phi + \cot^4\phi)}. \qquad (16)$$

All the stress components must satisfy the equilibrium equations which, in the absence of body forces, are

$$\begin{aligned}
&\frac{\partial\sigma_{rr}}{\partial r} + \frac{1}{r}\frac{\partial\sigma_{r\theta}}{\partial\theta} + \frac{\partial\sigma_{rz}}{\partial z} + \frac{\sigma_{rr} - \sigma_{\theta\theta}}{r} = 0, \\
&\frac{\partial\sigma_{r\theta}}{\partial r} + \frac{1}{r}\frac{\partial\sigma_{\theta\theta}}{\partial\theta} + \frac{\partial\sigma_{\theta z}}{\partial z} + 2\frac{\sigma_{r\theta}}{r} = 0, \qquad (17) \\
&\frac{\partial\sigma_{rz}}{\partial r} + \frac{1}{r}\frac{\partial\sigma_{\theta z}}{\partial\theta} + \frac{\partial\sigma_{zz}}{\partial z} + \frac{\sigma_{rz}}{r} = 0.
\end{aligned}$$

where $\sigma_{rr} \equiv \sigma_{11}$, $\sigma_{r\theta} \equiv \sigma_{12}$, etc.

For any kinematically admissible deformation, we can determine the relevant values for S, $s_{r\theta}$ and s_{rz} from the constitutive equations. Hence, using (15), the equilibrium equations provide three first order partial differential equations for p, T_a and T_b. The solutions, which are straightforward in the cases of axial symmetry and pure bending, introduce arbitrary functions of integration, which are determined from the specified (admissible) boundary conditions.

1.4 Constitutive equations

The theory of constitutive equations for ideal-fibre-reinforced materials has already been treated in detail elsewhere (Chapter I). The general forms given there will simplify in the present case.

For *linear elastic behaviour*, the components of $\underset{\sim}{s}$ are related to those of $\underset{\sim}{e}$ through four material constants [2]; if the two families of fibres are mechanically equivalent (i.e. indistinguishable except for their directions) the number of constants reduces to three, as derived in Chapter I. Substitution of $\underset{\sim}{a}$ and $\underset{\sim}{b}$ from (2) into equation (49) of Chapter I produces fairly complicated relations between μ, γ_2, γ_7 and ϕ. It can be shown from these relations, or deduced by physical argument, that for balanced angle-ply reinforcement the stress-strain relations take the very simple form:

$$
\begin{aligned}
s_{11} &= Ee_{11}, & s_{22} &= Ee_{22}, & s_{33} &= Ee_{33}, \\
s_{12} &= 2G_{12}e_{12}, & s_{13} &= 2G_{13}e_{13}, & s_{23} &= 0.
\end{aligned}
\qquad (18)
$$

The material constants G_{12} and G_{13} may be interpreted as shear moduli, whilst E is related to the extensional modulus E_3 for uniaxial tension in the x_3-direction. We consider a rectangular block of balanced angle-ply IFRM with the two families of straight parallel fibres all lying in the planes x_1 = constant and such that the x_2 and x_3 axes bisect the fibre-directions as described by (2). Simple shear γ in the x_2-direction on the planes containing the fibres is defined by

$$\dot{\underset{\sim}{u}} = (0,\ 2\gamma x_1,\ 0);$$

from (15) and (18) we see that such a homogeneous deformation is supported by the stress field:

$$\sigma_{12} = 2G_{12}\gamma, \qquad \text{all other } \sigma_{ij} = 0$$

showing that G_{12} is the relevant shear modulus; similarly G_{13} is the shear modulus for simple shear in the x_3-direction. Uniform extension under simple tension in the x_3-direction is defined by (6) with e constant. Then

$$S = Ee,$$

and the condition that σ_{33} is the only non-zero stress component gives

$$\sigma_{33} = E_3 e, \qquad \text{all other } \sigma_{ij} = 0,$$

where the extensional modulus E_3 is related to E through

$$E_3 = 2(1 - \cot^2 \phi + \cot^4 \phi) E. \tag{19}$$

The Young's modulus for simple extension along the other bisector (the x_2-direction) is

$$E_2 = t^4 E_3. \tag{20}$$

The requirement that the strain energy function be a positive definite function of the strain components results in the simple conditions

$$G_{12} \geqslant 0, \qquad G_{13} \geqslant 0, \qquad E_3 \geqslant 0.$$

To describe *plastic-elastic response* of an IFRM we require a yield function f (σ_{ij}) and a flow rule. These are treated in detail elsewhere (Section 3 of Chapter I). For simplicity we assume non-hardening and adopt a quadratic (Mises-type) yield function, which for our balanced angle-ply cylinder takes the form (from equation (91) in Chapter I with (16) above):

$$f = \frac{s_{zz}^2}{Y^2} + \frac{s_{r\theta}^2}{k_2^2} + \frac{s_{rz}^2}{k_3^2} - 1 . \tag{21}$$

Here k_2 and k_3 represent shear yield stresses for axial and circumferential shear respectively on surfaces r = constant. They correspond to the values k_2 and k_1 respectively as defined in Chapter I. Y also differs from that given in Chapter I, and is now related to the yield stress Y_3 in simple axial tension through

$$Y_3 = 2(1 - \cot^2 \phi + \cot^4 \phi) Y. \tag{22}$$

For non-hardening materials, all four yield stresses are constants.

In considering only small elastic-plastic deformations, we make the usual assumption that the strain-rate $\underset{\sim}{d}$ can be expressed as the sum of two parts called the elastic and plastic strain-rates and denoted $\tilde{d}^e$ and $\tilde{d}^p$ respectively (refer Section 3 of Chapter I). Thus

$$d_{ij} = \frac{1}{2}\left(\frac{\partial v_i}{\partial x_j} + \frac{\partial v_j}{\partial x_i}\right) = d^e_{ij} + d^p_{ij}\,; \tag{23}$$

here $\tilde{v}$ represents the velocity of a particle at any time where, as is usual for quasi-static plastic deformation, *time* represents any convenient parameter which determines the sequence of events. The components d^e_{ij} of the elastic strain-rate are assumed to be linear functions of the components $\dot{s}_{ij}$ of an extra-stress rate. For sufficiently small deformations the equations must reduce to the relevant equations of linear elasticity, which are (18) in our case. The components d^p_{ij} of the plastic strain-rate are assumed to be associated with the yield function f (again refer Section 3, Chapter I for details); for balanced angle-ply materials they take a very simple form and, with (18), result in the following complete stress-strain relations:

$$d_{ij} = \frac{2\lambda}{YY_3} s_{ij} + \frac{1}{E}\dot{s}_{ij} \qquad (i = j), \tag{24}$$

$$d_{1\alpha} = \frac{\lambda}{k^2_\alpha} s_{1\alpha} + \frac{1}{2G_{1\alpha}}\dot{s}_{1\alpha} \qquad (\alpha = 2,3), \tag{25}$$

$$d_{23} = 0,$$

with

$$d_{11}:d_{22}:d_{33} = s_{11}:s_{22}:s_{33} = 1-t^2;-1:t^2. \tag{26}$$

Here λ is an arbitrary non-negative scalar multiplier and when $\lambda = 0$ the response is purely elastic.

Together with the yield condition (21), equations (24) and (25) yield four independent equations for the four unknowns λ, $s_{r\theta}$, s_{rz} and S ($= s_{zz}$) as functions of time and position, for any admissible deformation. They are highly non-linear equations. However, the analysis is considerably simplified when either $d_{r\theta} = s_{r\theta} = 0$ (as in some axisymmetric problems) or $d_{rz} = s_{rz} = 0$ (as in pure bending).

2. PROBLEMS WITH AXIAL SYMMETRY

2.1 A simple example - uniform inflation and axial extension

The preceding theory is simplified in the case of problems with axial symmetry. In order to illustrate the simplicity of the IFRM theory when applied to cylinders of highly anisotropic material, we first consider an example in which the analysis becomes elementary. The internal and external pressures are assumed to be uniform and we look for axially symmetric deformations for which the radial displacement is purely radial:

$$u = u(r). \tag{27}$$

Then (8) shows that such a displacement corresponds to

$$\psi(r,z) = \tfrac{1}{2}t^2z^2u/r \tag{28}$$

where, from (9), u satisfies the ordinary differential equation

$$\frac{d}{dr}\left(\frac{u}{r}\right) + \frac{2-t^2}{r}\,\frac{u}{r} = 0$$

giving solutions

$$u = Cr^{t^2-1}, \qquad v = 0, \qquad w = -t^2Czr^{t^2-2}, \tag{29}$$

where C is an arbitrary constant. The matrix of infinitesimal strain components is then readily given as

$$\underset{\sim}{e} \equiv \begin{pmatrix} e_{rr} & e_{r\theta} & e_{rz} \\ e_{r\theta} & e_{\theta\theta} & e_{\theta z} \\ e_{rz} & e_{\theta z} & e_{zz} \end{pmatrix} = Cr^{t^2-2}\begin{pmatrix} t^2-1 & 0 & \frac{1}{2}(2-t^2)z/r \\ 0 & 1 & 0 \\ \frac{1}{2}(2-t^2)z/r & 0 & -t^2 \end{pmatrix}. \tag{30}$$

We immediately see from (29) that uniform inflation can take place only if accompanied by axial extension (or compression), and that this extension is uniform (w independent of r) only if

$$\tan\phi = \surd 2. \tag{31}$$

This angle $\tan^{-1}\surd 2$ ($\sim 54^0 44'$) often arises in studies of this configuration. It is frequently used in filament winding practice and with reinforced hosepipes. The argument for this choice of winding angle is given by Young [3], and is based on the result that $\sigma_{\theta\theta} = 2\sigma_{zz}$ in a thin-walled helically reinforced closed cylindrical vessel under pressure. Maximum strength is therefore obtained when the fibre tensions contribute to $\sigma_{\theta\theta}$ and σ_{zz} in the ratio 2 : 1, and the requirement for this is that $t^2 = 2$.

The arguments in favour of using this angle for *thick-walled* cylinders under pressure are different, and follow from the observation that the shear strain component e_{rz} is not zero except for $t = \sqrt{2}$. On *any* reasonable model of physical behaviour (other than that of a perfect fluid), a non-zero shear strain must be associated with a corresponding non-zero shear extra-stress component. Hence since the shear component of traction at both the lateral surfaces is prescribed to be zero, the shear stress σ_{rz} is discontinuous across each of these surfaces. As discussed elsewhere (see Chapter II) a discontinuity gives rise to a singular sheet of fibres at the surface, and in real materials (with small fibre-inextensibilities) it should be interpreted as a rapidly varying shear stress causing high concentrations of the stress components $\sigma_{\theta\theta}$ and σ_{zz} in a thin layer at the surface (refer Chapter IV). The actual magnitudes of the stress singularities will be discussed in the next sub-section 2.2; for the present we simply note that in the design of components such stress concentrations must be undesirable, and that they do *not* arise in the case of $t = \sqrt{2}$.

Assuming this special angle of winding, we find that the deformation takes a very simple form, namely

$$\begin{pmatrix} u \\ v \\ w \end{pmatrix} = C \begin{pmatrix} r \\ 0 \\ -2z \end{pmatrix}, \qquad \underset{\sim}{e} = C \begin{pmatrix} 1 & 0 & 0 \\ 0 & 1 & 0 \\ 0 & 0 & -2 \end{pmatrix}. \tag{32}$$

From (14), (15) and (18), the stress components are given by

$$\sigma_{rr} = -\tfrac{1}{2}S - p, \qquad \sigma_{\theta\theta} = -\tfrac{1}{2}S - p + \tfrac{2}{3}(T_a + T_b),$$

$$\sigma_{zz} = S - p + \tfrac{1}{3}(T_a + T_b), \qquad \sigma_{\theta z} = \frac{\sqrt{2}}{3}(T_a - T_b), \qquad \sigma_{rz} = \sigma_{r\theta} = 0, \tag{33}$$

where we have also incorporated the assumption that a zero shear strain is associated with a zero shear extra-stress component.

Axial symmetry and the end condition on $\sigma_{\theta z}$ then imply that

$$T_a = T_b = T, \text{ say.} \tag{34}$$

Note that in fact the symmetry of the general axisymmetric problem, not just this special case, requires that $T_a = T_b$, so for the remainder of this chapter we refer only to T.

We further note that S depends only on the strain components, and hence is a function only of C; so S is independent of position in the cylinder. Thus the equilibrium equations (17) finally reduce to

$$r\frac{\partial p}{\partial r} + \tfrac{4}{3}T = 0, \qquad \frac{\partial}{\partial z}(p - \tfrac{2}{3}T) = 0\,. \tag{35}$$

The solution is elementary:

$$p = \frac{1}{r^2}\left\{2\int_{r_0}^{r} \eta g(\eta)\, d\eta + \ell(z)\right\}, \tag{36}$$

$$T = \frac{3}{2r^2}\left\{-\int_{r_0}^{r} \eta^2 g'(\eta)\, d\eta + \ell(z)\right\}. \tag{37}$$

With (33), these expressions provide the stress solution. The two arbitrary functions $g(r)$ and $\ell(z)$, together with C, can be determined from the boundary conditions (1) provided the assumed displacement field (32) is compatible with these conditions. Thus the end condition on σ_{zz} gives

$$g(r) = p - \tfrac{2}{3}T = S - N(r), \tag{38}$$

and the pressure condition on the internal boundary then yields

$$\ell(z) = (p_0 - \tfrac{3}{2}S)r_0^2. \tag{39}$$

Finally we obtain C implicitly through S, provided there is a one-to-one correspondence; S is given by applying the pressure condition on the external boundary. Thus

$$S = \frac{2\{p_1 r_1^2 - p_0 r_0^2 + 2F(r_1)\}}{3(r_1^2 - r_0^2)} \tag{40}$$

where

$$F(r) = \int_{r_0}^{r} \eta N(\eta)\, d\eta, \tag{41}$$

so that the complete stress solution for uniform pressures p_0, p_1 is

$$\sigma_{rr} = \frac{2}{r^2}\left\{F(r) - \frac{r^2 - r_0^2}{r_1^2 - r_0^2} F(r_1)\right\} - p_0 \frac{r_0^2}{r^2}\frac{r_1^2 - r^2}{r_1^2 - r_0^2} - p_1 \frac{r_1^2}{r^2}\frac{r^2 - r_0^2}{r_1^2 - r_0^2},$$

$$\sigma_{\theta\theta} = -\frac{2}{r^2}\left\{F(r) + \frac{r^2 + r_0^2}{r_1^2 - r_0^2} F(r_1)\right\} + p_0 \frac{r_0^2}{r^2}\frac{r_1^2 + r^2}{r_1^2 - r_0^2} - p_1 \frac{r_1^2}{r^2}\frac{r^2 + r_0^2}{r_1^2 - r_0^2} + 2N(r), \tag{42}$$

$$\sigma_{zz} = N(r), \quad \sigma_{r\theta} = \sigma_{rz} = \sigma_{\theta z} = 0.$$

It is interesting to note that in this special case of $\phi = \tan^{-1}\sqrt{2}$ the stress field is completely independent of the constitutive equation; only the magnitude of the deformation, governed by C, is affected. This obviously suggests that one way of deducing the constitutive behaviour of such a balanced angle-ply material is to measure the deformation produced by specified boundary fractions.

For linear elastic response, S is proportional to C:

$$S = Ee_{zz} = -2EC,$$

where, from (19), E is related to the extensional moduli by

$$E = \tfrac{1}{6}E_2 = \tfrac{2}{3}E_3 .$$

Hence

$$C = \frac{p_0 r_0^2 - p_1 r_1^2 - 2F(r_1)}{3E(r_1^2 - r_0^2)} . \tag{43}$$

For plastic-elastic response, we see from (21) that yielding ($f = 0$) occurs simultaneously at all points of the cylinder when $S = \pm Y$, i.e. when

$$C = \pm\tfrac{1}{2}Y/E = \pm Y_3/E_3 , \tag{44}$$

or equivalently when the boundary tractions become such that

$$p_0 r_0^2 - p_1 r_1^2 - 2F(r_1) = \pm\tfrac{3}{2}(r_1^2 - r_0^2)Y. \tag{45}$$

For definiteness we consider the case in which C is positive and increases from zero. Then for $C < Y_3/E_3$ the elastic solution applies. For $C \geqslant Y_3/E_3$, the value of S remains constant (= -Y) and the relation between S and C is no longer one-to-one; the deformation is still given by (32) and for the solution to be valid the boundary tractions must continue to satisfy the condition (45).

The resultant axial force applied to the ends of the cylinder is $2\pi F(r_1)$. Hence the condition (45) shows that the internal pressure at which the cylinder yields is increased by the application of a tensile axial force. It also shows that if the cylinder is subject to the *closed-end* condition in the form $2\pi F(r_1) = \pi r_0^2 p_0 - \pi r_1^2 p_1$, then plastic yielding can never occur, whatever the magnitude of the tractions; indeed, (40) shows that then S = 0, so zero deformation is predicted.

2.2 Singular stress layers

This example illustrates the simplicity introduced into the analysis by using IFRM theory. However, it *is* very special, as indeed it was chosen to be, in order that singular stress layers should not occur. The analysis shows that if

(i) the inflation of the cylinder is purely radial, and

(ii) $\tan\phi = \sqrt{2}$,

the deformation reduces to one in which all the shear strain components are zero; then the boundary conditions of zero shear stress are automatically satisfied (at least for most material behaviours).

From a design point of view, the implications of the results are good. The absence of shear strain and stress reduces the possibility of failure by a delamination mechanism. The deformation also enables the axial load and hoop stress to be quite uniformly distributed through the thickness of the cylinder, thus making effective use of all the reinforcing fibres. But the most important consequence is the absence of singular stress layers and their implication of possible fracture.

However, if $t \neq \sqrt{2}$ *or* if the radial displacement depends on z as well as r, then these implications all disappear. Thus (30) shows that in the former case ($t \neq \sqrt{2}$) a purely radial inflation produces a non-uniform strain field in which $e_{rz} = -C(t^2-2)zr^{t^2-3}$, so $e_{rz} \neq 0$ except at $z = 0$ or when $C = 0$ (no deformation). If the inflation is not purely radial, so that $u = u(r,z)$, then even with axial symmetry and $t = \sqrt{2}$ the simplest displacement field (from (7)) is

$$u = rh'(z), \qquad v = 0, \qquad w = -2h(z), \tag{46}$$

where h(z) is an arbitrary function; this deformation leads again to a non-uniform strain field in which now $e_{rz} = \frac{1}{2}rh''(z)$, so $e_{rz} \neq 0$ on the lateral boundaries $r = r_0$ and $r = r_1$ (except in the special case of $u = Cr$). Hence in both cases the corresponding shear stress σ_{rz} ($= s_{rz}$) is non-zero, so that for the solutions to be valid the tractions on the lateral boundaries should have a very particular dependence on z and those on the ends a very particular dependence on r. Otherwise there is a mismatch between the stresses and the specified boundary values, leading to boundary layers of singular in-surface stress.

These are just two instances of cylinder problems involving singular stress layers. Their occurrence is not restricted to the context of axial symmetry - they arise in nearly all boundary value problems, and one of the principal advantages of the IFRM model is that the magnitude of the singularity can be readily obtained.

The equilibrium equations (17) show that a discontinuity in shear stress σ_{rz} can be sustained only by singular direct stresses associated with the hydrostatic pressure and fibre tensions. As in plane strain (discussed elsewhere) these singularities are represented by p and/or T including Dirac delta functions; physically this corresponds to a thin sheet carrying a finite *force* in its surface. Equations (17) also show that p and T may be singular even *without* a discontinuity in shear; this corresponds to a fibre surface sustaining a normal pressure and surface tension without change in shape or length and might occur, for example, if point loads were applied on a boundary.

In the following we demonstrate the relevant theory for axially symmetric problems, and show that the singular layers can take the form of cylindrical layers (r = constant) or normal cross-sectional surface layers (z = constant).

Symmetry implies that the equilibrium equation (17), with (14) and (15), yield two first-order differential equations for p and T in terms of σ_{rz} (= s_{rz}) and S (= s_{zz}):

$$\frac{\partial p}{\partial r} + 2\sin^2\phi\frac{T}{r} = \frac{\partial\sigma_{rz}}{\partial z} + (\cot^2\phi - 1)\frac{\partial S}{\partial r} + (2\cot^2\phi - 1)\frac{S}{r},$$
$$\frac{\partial}{\partial z}(p - 2T\cos^2\phi) = \frac{1}{r}\frac{\partial}{\partial r}(r\sigma_{rz}) + \frac{\partial S}{\partial z}. \tag{47}$$

Cylindrical layers. A jump of $\tau_r(z)$ in σ_{rz} across a cylindrical surface $r = \rho$ is associated with singularities in T at $r = \rho$. This can be seen from (47). The second equation shows that $p - 2T\cos^2\phi$ must be singular; however, the first shows that p cannot be singular since otherwise this would produce a term involving the derivative $\delta'(r-\rho)$ which could not be balanced in the equation.

Accordingly we assume that at $r = \rho$

$$T = \tfrac{1}{2}\sec^2\phi\, F_\rho(z)\,\delta(r-\rho). \tag{48}$$

The $\sec^2\phi$ term has been introduced so that $F_\rho(z)$ should conveniently denote the *force* per unit length which acts in the z-direction on cross-sectional circles in the surface $r = \rho$; the force acting in the circumferential direction is then t^2F_ρ.

Now consider the equilibrium of a thin cylindrical shell which includes the surface $r = \rho$. The equilibrium conditions for this are conveniently obtained by integrating (47) through the surface $r = \rho$; this gives

$$\frac{t^2}{\rho}F_\rho = -[p] + (\cot^2\phi - 1)[S]$$
$$-\frac{dF_\rho}{dz} = [\sigma_{rz}] = \tau_r(z), \tag{49}$$

where [p], for example, denotes the jump in p in going from $r = \rho-0$ to $r = \rho+0$. Hence

$$F_\rho = F_\rho(L) + \int_z^L \tau_r(z')\, dz' \tag{50}$$

with an associated jump in p of

$$[p] = -t^2 F_\rho/\rho + (\cot^2\phi - 1)[S] \,. \tag{51}$$

Thus at $r = \rho$ the stress field has the properties

$$[\sigma_{rr}] = -[p] + [s_{rr}] = \left(\frac{1}{2} - \frac{1}{t^2}\right)[S] + t^2 F_\rho/\rho \,, \qquad [\sigma_{rz}] = \tau_r(z) \tag{52}$$

and

$$\sigma_{\theta\theta} = \rho[\sigma_{rr}]\delta(r-\rho) \,, \qquad \sigma_{zz} = F_\rho(z)\delta(r-\rho) \,, \tag{53}$$

where F_ρ is given by (50).

When $r = \rho$ is an interior surface, the jumps in σ_{rr} and σ_{rz} are determined by the deformation and constitutive equations of the material on each side of $r = \rho$; when it is a boundary surface, the values of σ_{rr} and σ_{rz} on the material side are still determined in that way, but the values on the other side are now specified boundary values.

Singular normal layers. Discontinuous σ_{rz} across any plane cross-section $z = z_0$ results in infinite values of p as well as T, as can be seen from (47). Equilibrium is maintained by finite loads being carried in the cross-section, with the radial and circumferential stresses being singular but the axial stress remaining continuous.

Thus, for a jump of $\tau_z(r)$ in σ_{rz} across $z = z_0$, we write

$$p = P(r)\delta(z-z_0) \,, \qquad T = R(r)\delta(z-z_0) \,. \tag{54}$$

Then the equilibrium equations (47) imply that

$$\frac{dP}{dr} + 2\sin^2\phi\,\frac{R}{r} = \tau_z(r)$$

and

$$P - 2\cos^2\phi\, R = 0 \,.$$

Hence the radial force in the surface $z = z_0$ is given by

$$P = 2R\cos^2\phi = r^{-t^2}\int^{r} r^{t^2}\tau_z(r)\,dr\,. \tag{55}$$

This relation shows that any ring pressure concentrated *loads* P_0, P_1 per unit length, applied at $r = r_0$ and $r = r_1$, together with the jump $\tau_z(r)$, must satisfy the integral condition

$$P_1 r_1^{t^2} - P_0 r_0^{t^2} = \int_{r_0}^{r_1} r^{t^2}\tau_z(r)\,dr\,. \tag{56}$$

This also shows that a non-zero force can be supported in such a singular stress layer even if no ring loads are applied on the lateral boundaries. And, lastly, we note that the presence of a singular in-plane stress in a normal surface still leaves σ_{zz} continuous across the surface (though there *is* an instantaneous jump in the total fibre load).

Since the inextensible fibre-layers (r = constant) can support finite loads, we expect that singular cylindrical layers can occur not only in IFRM cylinders but also in those cylinders which consist of *compressible* material reinforced with inextensible fibres. This is not the case for the normal singular surfaces, which can be supported only by the incompressibility of IFRM. However, the normal layers are still very useful for real fibre-reinforced materials with small compressibility, and their occurrence could presage failure by interlaminar debonding.

2.3 Example - clamped pressurised cylinder

Short reinforced cylinders are often used to connect cylindrical pipes carrying pressurised liquid. We consider the same configuration as before, except that now the two ends $z = \pm L$ are bonded onto rigid circular rings of radius r_0 (Figure 2) but are otherwise unconstrained; hence the end conditions are now

$$\sigma_{zz} = N(r) = 0 \qquad \text{for} \qquad z = \pm L\,, \tag{57}$$

$$u = 0 \qquad \text{for } r = r_0,\ z = \pm L\,. \tag{58}$$

For simplicity we assume zero external pressure

$$p_1 = 0, \qquad p_0 = \text{constant}$$

and the special winding angle

$$\phi = \tan^{-1}\sqrt{2}\,.$$

We also assume linearly elastic behaviour, governed by the constitutive equations (18).

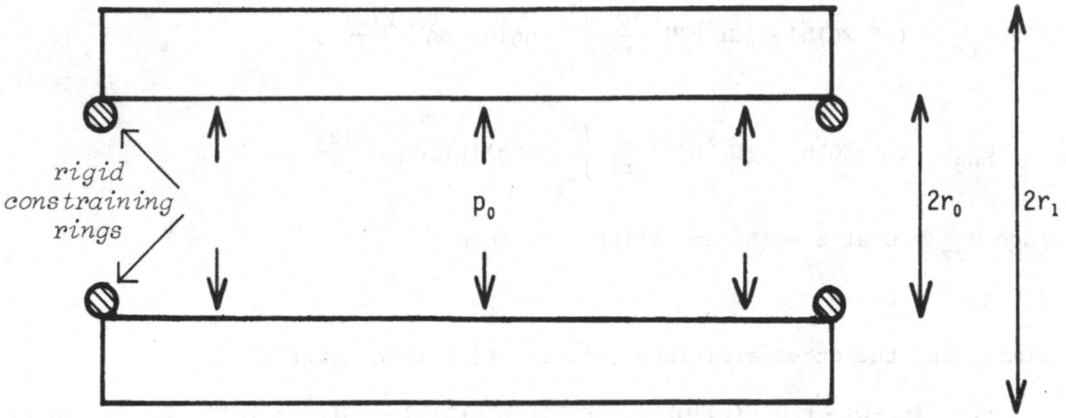

Figure 2 Axial cross-section of pressurised cylinder with end constraints

The axial symmetry, and $t^2 = 2$, reduces equation (9) to simply $\partial^3\psi/\partial r\partial z^2 = 0$ so that, with symmetry with respect to $z = 0$, the solution (8) takes the form

$$u = rh'(z), \qquad v = 0, \qquad w = -2h(z), \tag{59}$$

where $h' \equiv dh/dz$, and $h(z) = h(-z)$. Hence, from (58), we immediately see that no radial displacement takes place at the two ends, with

$$h'(L) = 0. \tag{60}$$

From (59), and using (18), we obtain

$$\underset{\sim}{e} = \begin{pmatrix} h' & 0 & \tfrac{1}{2}rh'' \\ 0 & h' & 0 \\ \tfrac{1}{2}rh'' & 0 & -2h' \end{pmatrix}, \qquad \underset{\sim}{s} = \begin{pmatrix} Eh' & 0 & Grh'' \\ 0 & Eh' & 0 \\ Grh'' & 0 & -2Eh' \end{pmatrix}, \tag{61}$$

where $G \equiv G_{13}$. Then the equilibrium equations (47) reduce to

$$\frac{\partial p}{\partial r} + \frac{4}{3}\frac{T}{r} = Grh'' ,$$

$$\frac{\partial}{\partial z}(p - \tfrac{2}{3}T) = 2(G-E)h'' . \tag{62}$$

These are easily solved for p and T, leading to the stress solution

$$\sigma_{zz} = -2Gh'(z) + g(r), \qquad \sigma_{rz} = Grh'', \qquad \sigma_{\theta z} = 0 = \sigma_{r\theta},$$

$$\sigma_{rr} = (3E-2G)h' - \tfrac{1}{4}Gr^2h''' - \frac{2}{r^2}\int_{r_0}^{r} \eta g(\eta)\, d\eta - \frac{\ell(z)}{r^2} , \tag{63}$$

$$\sigma_{\theta\theta} = (3E-2G)h' + \tfrac{1}{4}Gr^2h''' - \frac{2}{r^2}\int_{r_0}^{r} \eta^2 g'(\eta)\, d\eta + \frac{\ell(z)}{r^2} .$$

Since $\sigma_{zz} = 0$ at $z = \pm L$, and $h'(L) = 0$, then

$$g(r) = 0.$$

Eliminating the other arbitrary function $\ell(z)$ then gives

$$r_1^2\sigma_{rr}(r_1-0) - r_0^2\sigma_{rr}(r_0+0) = (3E-2G)(r_1^2-r_0^2)h' - \tfrac{1}{4}G(r_1^4-r_0^4)h''' . \tag{64}$$

The discontinuity in σ_{rz} across the curved boundary surfaces $r = r_0$ and $r = r_1$ imply singular stress layers there; their magnitudes $F_0(z)\delta(r-r_0)$ and $F_1(z)\delta(r-r_1)$ are determined from (50) as

$$F_0 = -Gr_0h' , \qquad F_1 = Gr_1h' . \tag{65}$$

The associated jumps in σ_{rr}, from (52), thus yield

$$\sigma_{rr}(r_0+0) = -p_0 - 2Gh' , \qquad \sigma_{rr}(r_1-0) = -2Gh' ,$$

so that, from (64), we find that h(z) must satisfy

$$\tfrac{1}{4}G(r_1^4-r_0^4)h''' - 3E(r_1^2-r_0^2)h' = -p_0r_0^2 \tag{66}$$

for $|z| < L$. The solution of this which satisfies $h'(L) = 0$ and the symmetry conditions

$$\sigma_{rz} = 0, \quad w = 0 \qquad \text{for } z = 0, \qquad r_0 < r < r_1,$$

is

$$h(z) = \frac{r_0^2}{3(r_1^2-r_0^2)} \frac{p_0}{E}\left(z - \frac{\sinh bz}{b\cosh bL}\right), \tag{67}$$

where

$$b^2 = \frac{12E}{G(r_0^2+r_1^2)} > 0, \tag{68}$$

so that b is real. Hence

$$u = rh'(z) = \frac{rr_0^2}{3(r_1^2-r_0^2)} \frac{p_0}{E}\left(1 - \frac{\cosh bz}{\cosh bL}\right) \tag{69}$$

and

$$\sigma_{rz} = Grh''(z) = -\frac{brr_0^2 p_0}{3(r_1^2-r_0^2)} \frac{G\sinh bz}{E\cosh bL}, \qquad |z| < L. \tag{70}$$

We note that this non-uniform radial expansion has the property that, if $bL \gg 1$,

$$u(r,z) \sim u(r,0) = \frac{rr_0^2 p_0}{3(r_1^2-r_0^2)E},$$

agreeing with the value predicted by the solution (43) for uniform inflation. Thus for a long cylinder the end effects due to the constraining rings are confined to small regions in the neighbourhoods of $z = \pm L$, with exponential decay:

$$u(r,z) \sim u(r,0)\{1 - \exp b(|z|-L)\}.$$

The constraining rings have to exert radial tension $-P_0$ in order to maintain equilibrium. The magnitude of this force is given by (56) and (69) as

$$-P_0 = -\frac{1}{r_0^2}\int_{r_0}^{r_1} r^2\sigma_{rz}(r,L-0)\,dr = \tfrac{1}{12}(r_0^2+r_1^2)\frac{G}{E}bp\tanh bL.$$

Hence there is an obvious danger of delamination at the ends, due to the singular radial tensions there. Failure by fracture of the fibres is most likely to occur at the mid-plane $z = 0$, where the fibre load takes its maximum value at the outer surface $r = r_1$.

2.4 General axial symmetry

Although much of the analysis has been simplified by assuming $\tan\phi = \sqrt{2}$, the general theory for $t^2 \neq 2$ involves little added difficulty. The details can be found in [4]. The most general displacement field is shown to be

$$u = r^{t^2-1}h'(z), \qquad v = 0, \qquad w = t^2 r^{t^2-2}h(z) + k(r);$$

$k(r)$ describes a telescopic shear which is independent of z, and is associated with the anti-symmetric part of any problem, such as when the lateral surfaces are subjected to displacements or tractions which are *not* symmetric under reflection in $z = 0$.

The stress analysis is much the same as for $t = \sqrt{2}$, and is shown to lead, in general, to solutions of the form

$$h(z) = \alpha_1 \sinh b_1 z + \alpha_2 \sinh b_2 z + q(z),$$

where α_1 and α_2 are arbitrary constants, $q(z)$ is associated only with the boundary pressures $p_0(z)$ and $p_1(z)$, and b_1 and b_2 are generalised forms of the constant b defined in (68).

In classical elasticity, solutions obtained for problems when the applied tractions are concentrated point or line loads are sometimes termed fundamental solutions. Solutions for more general boundary conditions are then obtained by superposition. The same can be done for fibre-reinforced materials, though in this case (as shown in the previous subsection 2.2) the infinite boundary tractions are not immediately dispersed from the points or lines of application, but are propagated through the body. The two basic cases of singular axial loading and singular radial loading are treated in [4].

The problem of inflation of a cylinder under uniform internal pressure and without end constraints is found to lead to non-uniform radial expansion, provided $t \neq \sqrt{2}$. Also, as expected, the lateral surfaces are singular stress layers, as are the plane ends.

3. PURE BENDING OF HOLLOW CYLINDERS

3.1 Kinematics

The kinematical constraints of IFRM are very restrictive and imply that many deformations which are of interest in unconstrained, or merely incompressible, materials are difficult to produce in highly anisotropic materials. Thus, for example, the conditions (7) do not allow pure torsion, nor uniform extension (unless $t^2 = 2$) nor, as we have seen, uniform inflation (unless $t^2 = 2$) nor, indeed, any but the simplest plane strain deformations. It is therefore remarkable that a solution of (7) is

$$u = -\tfrac{1}{2}R^{-1}\{(1-\cot^2\phi)r^2 + z^2\}\sin\theta,$$
$$v = -\tfrac{1}{2}R^{-1}\{(1-3\cot^2\phi)r^2 + z^2\}\cos\theta, \qquad (71)$$
$$w = R^{-1}rz\sin\theta.$$

This deformation is one of pure bending, in which the axis of the cylinder is bent so that it has a constant radius of curvature R. As in pure bending of an isotropic elastic solid, plane sections remain plane. The solution is independent of the constitutive equations.

The strains are given by (6) with

$$e = e_{zz} = R^{-1}r\sin\theta, \qquad e_{rz} = e_{\theta z} = 0 \quad \text{and}$$
$$e_{r\theta} = R^{-1}(2\cot^2\phi - 1)r\cos\theta. \qquad (72)$$

So this is yet another solution in which at least one shear strain component cannot be zero within the material, unless $\tan\phi = \sqrt{2}$, and therefore that in general there will be a mismatch in shear stress (in this case $\sigma_{r\theta}$) at any boundary at which that stress is specified. This will again lead to singular stress layers.

3.2 Boundary conditions

The ends of the cylinder ($z = \pm L$) are assumed to be subject to equal and opposite couples with moment M. The lateral boundaries ($r = r_0$,

$r = r_1$) are assumed free from traction and the resultant force on each of the ends is zero. Hence

$$\sigma_{rr} = \sigma_{r\theta} = 0 \qquad \text{for} \quad r = r_0,\ r = r_1\ , \tag{73}$$

and

$$\int_0^{2\pi}\int_{r_0}^{r_1} \sigma_{zz} r\, dr\, d\theta = 0, \qquad \int_0^{2\pi}\int_{r_0}^{r_1} \sigma_{zz} r^2 \sin\theta\, dr\, d\theta = M. \tag{74}$$

As M increases, so the radius of curvature R of the axis decreases; accordingly, for plastic deformation, it is convenient to choose the 'time' $\bar{t}$ (or ordering parameter) as R^{-1}:

$$\bar{t} = 1/R. \tag{75}$$

3.3 Special case $\tan\phi = \sqrt{2}$

Again the solution simplifies considerably when $\tan^2\phi = 2$, and it has already been remarked that in this case $e_{r\theta} = 0$, and hence in general

$$\sigma_{r\theta} = 0$$

throughout the cylinder, so that no singular stresses arise. To satisfy the remaining boundary conditions on $r = r_0$ and $r = r_1$, we choose p to equal s_{rr} everywhere (so $\sigma_{rr} = 0$), and T to be such that $\sigma_{\theta\theta} = \sigma_{rr} = 0$ for equilibrium. Hence in this case the only non-zero stress component through the cylinder is σ_{zz}, with

$$\sigma_{zz} = \begin{cases} R^{-1}E_3 r\sin\theta & \text{elastic behaviour,} \\ \pm Y_3 & \text{plastic behaviour.} \end{cases} \tag{76}$$

For the elastic solution, substitution for σ_{zz} into (74) gives

$$M = E_3 I_y/R \tag{77}$$

where $I_y = \frac{1}{4}\pi(r_1^4 - r_0^4)$ is the second moment of area of the cross-section about the x-axis; that is, the moment is proportional to curvature just as in elementary bending theory.

Yielding occurs when $\sigma_{zz} = \pm Y_3$, so that the plastic-elastic interfaces are the two *planes*

$$y = r\sin\theta = \pm RY_3/E_3 = \pm y_c \qquad \text{(say)}.$$

Plastic deformation first occurs at $y = \pm r_1$ when

$$\bar{t} = \bar{t}_0 = R_0^{-1} = Y_3/E_3 r_1, \qquad M = M_0 = I_y Y_3/r_1. \tag{78}$$

Subsequently, as M increases, the interfaces move towards $y = \pm r_0$ with (74) and (76) giving

$$M = \tfrac{1}{2}E_3R^{-1}\{r_1^4(\theta_c - \tfrac{1}{4}\sin 4\theta_c) - \tfrac{1}{2}\pi r_0^4\} + \tfrac{4}{3}Y_3 r_1^3\cos^3\theta_c, \tag{79}$$

where $\theta_c = \sin^{-1}(R/R_0)$. The interface reaches $y = \pm r_0$ when R decreases to the value

$$R_1 = E_3 r_0/Y_3 = (r_0/r_1)R_0$$

after which the moment-curvature relation becomes

$$M = \tfrac{1}{2}E_3R^{-1}\{r_1^4(\theta_c - \tfrac{1}{4}\sin 4\theta_c) - r_0^4(\theta_d - \tfrac{1}{4}\sin 4\theta_d)\}$$
$$+ \tfrac{4}{3}Y_3(r_1^3\cos^3\theta_c - r_0^3\cos^3\theta_d), \tag{80}$$

where $\theta_d = \sin^{-1}(R/R_1)$. The cylinder becomes *fully plastic* as $y_c \to 0$; then $R \to 0$ and

$$M \to \tfrac{4}{3}Y_3(r_1^3 - r_0^3).$$

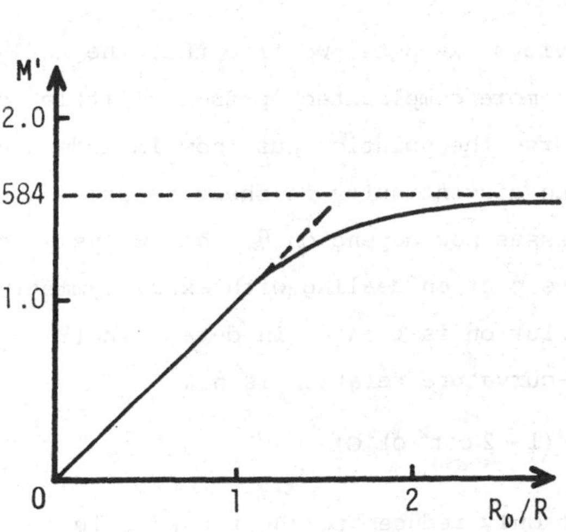

Figure 3 Dependence of bending moment on curvature

This is the maximum attainable value of M and may be regarded as the collapse load for the cylinder. The variation of M with R^{-1} is illustrated for $r_1/r_0 = 2$ in Figure 3, where a dimensionless moment $M' = r_1 M/Y_3 I_y$ is plotted against the normalised curvature R_0/R.

Finally we observe that, despite the complicated anisotropy of the configuration considered, in this special case of $t = \sqrt{2}$ the deformation is given in cartesian coordinates by

$$u_1 = -\frac{x_1 x_2}{2R}, \qquad u_2 = -\frac{1}{2R}\{\tfrac{1}{2}(x_2^2 - x_1^2) + x_3^2\}, \qquad u_3 = \frac{x_2 x_3}{R},$$

where $x_1 = r\cos\theta$, $x_2 = r\sin\theta$. This is exactly the same displacement field as that which arises in pure bending of an incompressible elastic-plastic *beam* of *isotropic* material, or of material which is transversely isotropic with respect to the axis of the beam. This correspondence in fact carries through to include the shape and development of the elastic-plastic interface, and the bending moment-curvature relation (provided, of course, that the material constants are suitably interpreted). Because of this correspondence in this special case of $\tan\phi = \sqrt{2}$ it is straight-forward to extend the analysis to include strain-hardening by, for example, a direct extension of the analysis given by Nadai [5].

3.4 General case $\tan\phi \neq \sqrt{2}$

As in the previous examples, we find that the analysis for general values of ϕ, though more complicated, presents little additional difficulty. Of course the solution must now include the singular stresses which arise with the discontinuity in shear stress $\sigma_{r\theta}$ across them. Although these stresses now depend on θ, the analysis is very similar to that of the previous section dealing with axial symmetry.

The elastic solution is treated in detail in [2]. It is found that the bending moment-curvature relation is now

$$M = R^{-1} I_y \{E_3 + (1 - 2\cot^2\phi)^2 G\}. \tag{81}$$

This expression not only reduces to the $t = \sqrt{2}$ solution of the previous section (and hence to the elementary bending result) but also shows that when $t \neq \sqrt{2}$ the bending moment required to produce a given curvature R^{-1} always exceeds the value suggested by elementary bending theory.

The plastic-elastic solution is described in [6]. There it is found that in general it is not possible to obtain an explicit expression for $\sigma_{r\theta}$ in the plastic region, unless $t = \sqrt{2}$. For this reason there is not an analytical expression relating the bending moment M to the curvature R^{-1}.

An interesting simple result concerns the shape of the plastic-elastic boundary, which we have seen takes the form of two parallel straight lines in the case $t = \sqrt{2}$. For the general case, (18), (21) and (72) show that *before* yield the Mises-type yield function takes the form

$$f = \left\{\frac{E_3}{RY_3} r \sin\theta\right\}^2 + \left\{(1 - 2\cot^2\theta)\frac{G}{Rk_2} r\cos\theta\right\}^2 - 1 . \tag{82}$$

For sufficiently small curvature (proportional to M, from (81)), this function is negative. At any given point (r,θ), f increases with M until it becomes zero when R is such that

$$\frac{r^2\cos^2\theta}{\alpha^2} + \frac{r^2\sin^2\theta}{\beta^2} = R^2 , \tag{83}$$

where

$$\alpha = |(1 - 2\cot^2\phi)^{-1}| k_2/G , \qquad \beta = Y_3/E_3 = Y/E . \tag{84}$$

Equation (83) represents an *ellipse* with semi-axes of length αR and βR along $\theta = 0$ and $\theta = \frac{1}{2}\pi$ respectively. In the case $t = \sqrt{2}$, $\alpha \to \infty$ and the ellipse degenerates into the two straight lines previously predicted.

We recall that, for $t = \sqrt{2}$, yield always occurs first on $r = r_1$ at the two points $(r_1, \pm\frac{1}{2}\pi)$. If $t \neq \sqrt{2}$, this happens only if $\alpha > \beta$, when $\bar{t} = \bar{t}_0$ and $R = R_0$ with

$$\bar{t}_0 = R_0^{-1} = Y/Er_1 ;$$

if $\alpha < \beta$, yield occurs first at $(r_1, 0)$ and (r_1, π).

As $\bar{t}$ increases from $\bar{t}_0$, the plastic region increases in size with the plastic-elastic interface moving inwards from the outer boundary. Since the extra-stress components in the elastic region remain unchanged, except for the variation in R^{-1}, the interface continues to be described by (83), with elastic behaviour inside the ellipse and plasticity outside it.

4. LARGE DEFORMATIONS OF THICK-WALLED CYLINDERS

4.1 Introduction

So far in this chapter, we have restricted attention to problems involving only small deformations. We now consider finite deformations of helically wound fibre-reinforced cylinders. There are obvious practical applications in the form of rubber and plastic tubes reinforced with strong fibres.

Although the kinematical conditions of incompressibility and fibre-inextensibility are severe constraints, it is still remarkable that analytical solutions to boundary value problems can be found in this context. The complexities of the non-linear theories of finite elasticity, viscoelasticity and plasticity are well-known, and there are still few such solutions available in these theories, even for isotropic materials. Indeed, it was because of this that the 'ideal' theory of IFRM was originally formulated.

In this section we present only a brief account of the basic results. Further details can be found in Spencer's book [7].

4.2 Kinematics and stress

We now have to make a distinction between quantities when referred to the undeformed configuration and those referring to the deformed configuration. We follow Chapter I in denoting the initial point by $\underset{\sim}{X}$ and its deformed position by $\underset{\sim}{x}$; in cylindrical polar coordinates, $\underset{\sim}{x}$ will still be described by components (r,θ,z) whilst $\underset{\sim}{X} \equiv (R,\Theta,Z)$. Components are referred to the base vectors of the undeformed configuration. We assume that in the reference configuration the winding angle for the two families of helices is $\pm\Phi$, and that $\pm\phi$ again denotes that angle after deformation. Of course, in general ϕ will now vary with position, whatever the dependence of Φ on R. For convenience we still assume Φ is constant throughout the cylinder, so that the initial directions of the fibres are given by the unit vectors

$$\underset{\sim}{a}^{(0)} = (0, \sin\Phi, \cos\Phi)\,, \qquad \underset{\sim}{b}^{(0)} = (0, -\sin\Phi, \cos\Phi)\,. \tag{85}$$

For axially symmetric deformations with uniform radial expansion we can still write

$$\underset{\sim}{a} = (0, \sin\phi, \cos\phi)\,, \qquad \underset{\sim}{b} = (0, -\sin\phi, \cos\phi)\,; \tag{86}$$

in general, we cannot.

In any deformation, the constraint of incompressibility requires

$$\det \underset{\sim}{F} = 1 \qquad \text{or equivalently} \qquad \det \underset{\sim}{C} = 1\,, \tag{87}$$

where $\underset{\sim}{F}$ is the deformation gradient ($F_{iA} = \partial x_i/\partial X_A$) and $\underset{\sim}{C} = \underset{\sim}{F}^T\underset{\sim}{F}$. Fibre-inextensibility in the directions $\underset{\sim}{a}$ and $\underset{\sim}{b}$ is satisfied provided (refer Chapter I, Section 2)

$$F_{iA}F_{iB}a_A^{(0)}a_B^{(0)} = C_{AB}a_A^{(0)}a_B^{(0)} = 1 \qquad (A,B = 1,2,3) \tag{88}$$

and similarly

$$C_{AB}b_A^{(0)}b_B^{(0)} = 1\,; \tag{89}$$

these follow from the relations

$$a_i = (\partial x_i/\partial X_A)a_A^{(0)} = F_{iA}a_A^{(0)}\,, \qquad b_i = F_{iA}b_A^{(0)}\,, \tag{90}$$

where both $\underset{\sim}{a}^{(0)}$ and $\underset{\sim}{a}$ are unit vectors (as are $\underset{\sim}{b}^{(0)}$ and $\underset{\sim}{b}$). In cylindrical polars the form of $\underset{\sim}{F}$ is

$$\underset{\sim}{F} = \begin{pmatrix} \dfrac{\partial r}{\partial R} & \dfrac{1}{R}\dfrac{\partial r}{\partial \Theta} & \dfrac{\partial r}{\partial Z} \\[2ex] r\dfrac{\partial \theta}{\partial R} & \dfrac{r}{R}\dfrac{\partial \theta}{\partial \Theta} & r\dfrac{\partial \theta}{\partial Z} \\[2ex] \dfrac{\partial z}{\partial R} & \dfrac{1}{R}\dfrac{\partial z}{\partial \Theta} & \dfrac{\partial z}{\partial Z} \end{pmatrix}.$$

Thus, for balanced angle-ply materials we find (from (85), (88) and (89)) that any kinematically admissible deformation must satisfy

$$C_{23} = 0\,, \qquad C_{22}\sin^2\Phi + C_{33}\cos^2\Phi = 1\,. \tag{91}$$

The incompressibility condition (86) does not reduce to such a convenient form, unless axial symmetry holds in which case we have

$$r = r(R,Z)\ , \qquad \theta = \Theta\ , \qquad z = z(R,Z)$$

so that the components of $\underset{\sim}{F}$ referred to the cylindrical polar axes are

$$\underset{\sim}{F} = \begin{pmatrix} \dfrac{\partial r}{\partial R} & 0 & \dfrac{\partial r}{\partial Z} \\ 0 & \dfrac{r}{R} & 0 \\ \dfrac{\partial z}{\partial R} & 0 & \dfrac{\partial z}{\partial Z} \end{pmatrix} . \tag{92}$$

Hence, from (90) we obtain (in an obvious notation)

$$\begin{aligned} a_r &= b_r = (\partial r/\partial Z)\cos\Phi\ , \qquad a_\theta = -b_\theta = (r/R)\sin\Phi\ , \\ a_z &= b_z = (\partial z/\partial Z)\cos\Phi\ ; \end{aligned} \tag{93}$$

$\underset{\sim}{a}$ and $\underset{\sim}{b}$ are unit vectors, so r and z must satisfy

$$\left(\frac{\partial r}{\partial Z} + \frac{\partial z}{\partial Z}\right)^2 \cos^2\Phi + \left(\frac{r}{R}\right)^2 \sin^2\Phi = 1\ . \tag{94}$$

Incompressibility yields the condition

$$\frac{\partial r}{\partial R}\frac{\partial z}{\partial Z} - \frac{\partial r}{\partial Z}\frac{\partial z}{\partial R} = 0\ . \tag{95}$$

Equations (87) - (89) provide three equations for the three position coordinates, and any solution will define a kinematically admissible deformation. As in the case of small deformation (and as discussed elsewhere (Chapter II)), such deformations are also statically admissible. For any admissible deformation, the extra-stress can be computed using the relevant constitutive equations and the stress solution is completed by then determining the arbitrary functions p, T_a and T_b from the equilibrium equations. Singular stress layers again arise, in general, and must be incorporated into the analysis.

For cylinder problems with axial symmetry, the theory governing stress and equilibrium is very similar to that just given in Sections 1 and 2 of Chapter V, with care being taken now to incorporate the possible dependence of ϕ on position.

4.3 Finite inflation and extension

To illustrate some of the relevant theory, we examine deformations of the form

$$r = r(R), \qquad \theta = \Theta, \qquad z = Zw(R),$$

which describes uniform radial inflation and axial extension. For this to be an admissible deformation, the functions $r(R)$ and $w(R)$ must satisfy (94) and (95), giving

$$\frac{r^2}{R^2}\sin^2\Phi + w^2\cos^2\Phi = 1$$

and

$$w\,\frac{r}{R}\,\frac{\partial r}{\partial R} = 1.$$

Eliminating w then yields

$$\frac{r}{R}\frac{dr}{dR} = \frac{\cos\Phi}{\{1-(r/R)^2\sin^2\Phi\}^{\frac{1}{2}}}. \tag{96}$$

The solution is most conveniently obtained parametrically in terms of the fibre angle ϕ after deformation. Equations (86) and (90) show that

$$\sin\phi = \frac{r}{R}\sin\Phi, \qquad \cos\phi = w\cos\Phi; \tag{97}$$

then (95) can be integrated to give

$$\log R = \int^{\phi} \frac{\cos^2\phi\sin\phi}{\sin^2\Phi\cos\Phi - \sin^2\phi\cos\phi}\,d\phi.$$

These give R, r and w all in terms of ϕ, and hence determine the entire deformation. This can then be used to obtain the stress solution, as described in the previous section; the rigid-plastic solution is straight-forward (refer [7]) though with complicated algebra.

REFERENCES

[1] LEKHNITSKII, S.G., *Theory of Elasticity of an Anisotropic Body*, Holden-Day, San Francisco, 1963

[2] SPENCER, A.J.M., MOSS, R.L. and ROGERS, T.G., *J. Elasticity* 5 (1975) 287-296

[3] YOUNG, R.E., *Proc. 13th Annual Meeting of the Reinforced Plastics Division, The Society of the Plastics Industry, Chicago*, 1958

[4] MOSS, R.L., ROGERS, T.G. and SPENCER, A.J.M., *J. Inst. Maths Applics* 26 (1980) 21-38

[5] NADAI, A., *Theory of Flow and Fracture of Solids*, McGraw-Hill, New York, 1950

[6] ROGERS, T.G., SPENCER, A.J.M. and MOSS, R.L., *Proc. 15th Int. Cong. Theor. & App. Mech., Toronto*, 1980

[7] SPENCER, A.J.M., *Deformations of Fibre-Reinforced Materials*, Clarendon Press, Oxford, 1972

[8] MULHERN, J.F., *Quart. J. Mech. Appl. Math.* 22 (1969) 97-114

VI

FRACTURE MECHANICS OF FIBER-REINFORCED MATERIALS

A. C. PIPKIN

Division of Applied Mathematics
Brown University
Providence, Rhode Island 02912
U.S.A.

1. Introduction

In plane stress of materials composed of two orthogonal families of strong fibers, the analysis based on the assumption of fiber inextensibility is so simple that it becomes possible to predict the course of an advancing crack or tear, and to determine how the shape of the crack depends on the failure criterion.

According to the idealized theory in which fibers are treated as continuously distributed and inextensible, an infinitesimally thin fiber can carry a finite force. This force is represented by a Dirac delta in the tensile stress. In a material composed of two orthogonal families of inextensible fibers, the fibers that pass through the tip of a crack are singular fibers of this kind. The forces in these fibers, evaluated at the crack tip, are closely related to the elastic stress intensity factors that characterize the stress singularity in an elastic analysis of the same problem.

In elastic fracture analysis it is assumed that the failure criterion for the material can be expressed in terms of the stress intensity factors. By using the relation between tip forces and stress intensity factors, we can express failure criteria in terms of the tip forces given by the inextensible theory.

In elastic fracture mechanics it seems to be an ingrained assumption that failure is explained by the Griffith criterion; so ingrained, in fact, that "failure criterion" is often used as a synonym for "energy release rate" even in experimental papers that demonstrate that they are not the same thing. For this reason, I will take a little time to explain what I mean by a failure criterion (Section 4).

Although the inextensible theory is only an approximation to elasticity theory, its loss of accuracy is more than compensated by the ease with which problems can be solved. It is possible to predict crack trajectories under a variety of assumptions about the failure criterion for the material, to see what failure criterion gives results that agree with the experimental facts.

Some quasi-static crack trajectories are computed in Section 5. Conditions governing dynamic crack propagation are sketched in Section 6. These are used in Section 7 to compute the speed of propagation of a straight crack, and in Section 8 to solve a quasi-static problem that has no static solution.

2. The inextensible theory.

The general kind of problem we wish to consider is illustrated in Figure 1. We consider plane stress of a sheet that is torn or cracked along a curve that runs from the outer boundary to a point (x_0,y_0) inside the sheet. With respect to an origin at the crack tip, the crack lies in the third quadrant, and we suppose that no line x = constant or y = constant crosses the crack more than once.

The sheet is composed of strong fibers that run parallel to the x and y directions. With the idealization that these fibers are continuously distributed and inextensible, the displacement components u and v satisfy

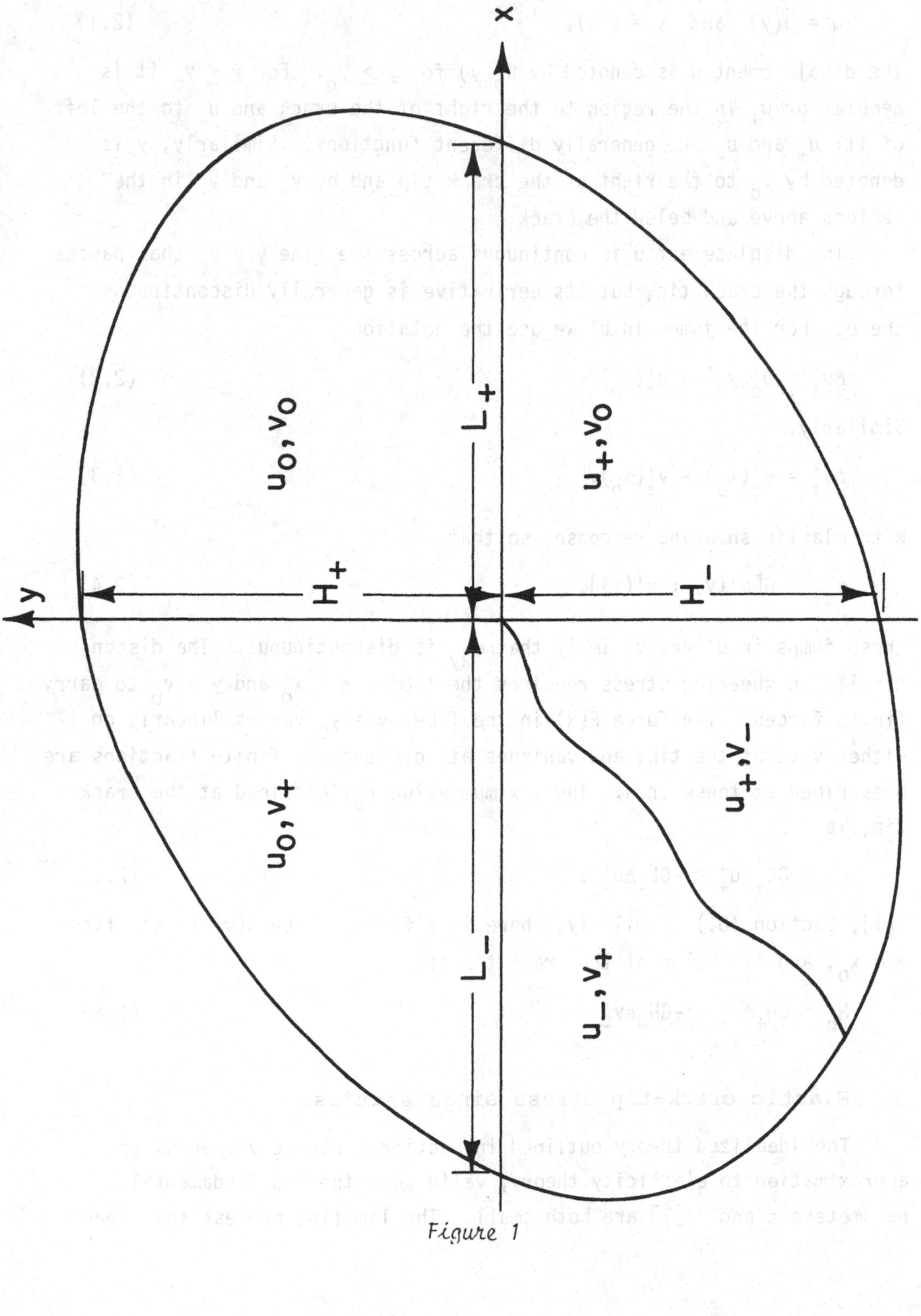

Figure 1

$$u = u(y) \quad \text{and} \quad v = v(x). \tag{2.1}$$

The displacement u is denoted by $u_0(y)$ for $y \geq y_0$. For $y \leq y_0$ it is denoted by u_+ in the region to the right of the crack and u_- to the left of it; u_+ and u_- are generally different functions. Similarly, v is denoted by v_0 to the right of the crack tip and by v_+ and v_- in the regions above and below the crack.

The displacement u is continuous across the line $y = y_0$ that passes through the crack tip, but its derivative is generally discontinuous there. For the jumps in u' we use the notation

$$\Delta u'_\pm = u'_0(y_0) - u'_\pm(y_0). \tag{2.2}$$

Similarly,

$$\Delta v'_\pm = v'_0(x_0) - v'_\pm(x_0). \tag{2.3}$$

With elastic shearing response, so that

$$\sigma_{xy} = G[u'(y) + v'(x)], \tag{2.4}$$

these jumps in u' and v' imply that σ_{xy} is discontinuous. The discontinuity in shearing stress requires the fibers $x = x_0$ and $y = y_0$ to carry finite forces. The force F(x) in the fiber $y = y_0$ varies linearly on either side of the tip, and vanishes at both ends if finite tractions are prescribed at these ends. The maximum value F_0, attained at the crack tip, is

$$F_0 = GL_+\Delta u'_+ = -GL_-\Delta u'_- . \tag{2.5}$$

([1], Section 10.) Similarly, there is a finite force N(y) in the fiber $x = x_0$, and its value at the crack tip is

$$N_0 = GH_+\Delta v'_+ = -GH_-\Delta v'_- . \tag{2.6}$$

3. Elastic crack-tip stress singularities.

The idealized theory outlined in Section 2 can be viewed as an approximation to elasticity theory, valid when the two fundamental parameters ε and c [1] are both small. The limiting process that leads to the idealized theory does not give the details of the stress distribution in the stress concentration layers that pass through the crack tip, but instead yields only the total tensile forces F(x) and N(y) in these layers.

In elasticity theory the stress components vary with distance r from a crack tip in proportion to $1/\sqrt{r}$. For a crack pointing toward the direction θ_o (i.e. along the ray $\theta = \theta_o+\pi$) with its tip at the origin, the Airy stress function is given to leading order (with respect to distance from the tip) by $\chi = K_u\chi_u + K_v\chi_v$ [2], where K_u and K_v are constants and

$$\chi_u = (4/3)\mathrm{Re}[i\varepsilon t_1(z_1/t_1)^{3/2} - it_2(z_2/t_2)^{3/2}] \tag{3.1}$$

and

$$\chi_v = (4/3)\mathrm{Re}[t_1(z_1/t_1)^{3/2} - ct_2(z_2/t_2)^{3/2}]. \tag{3.2}$$

Here

$$z_1 = x + icy, \quad z_2 = \varepsilon x + iy, \tag{3.3}$$

and

$$t_1 = \cos\theta_o + ic\sin\theta_o, \quad t_2 = \varepsilon\cos\theta_o + i\sin\theta_o. \tag{3.4}$$

We will set

$$K_u = A[(\sin\theta_o)/\varepsilon]^{1/2}, \quad K_v = B[(\cos\theta_o)/c]^{1/2}, \tag{3.5}$$

with no loss of generality unless $\sin\theta_o$ or $\cos\theta_o$ is zero.

In the theory of stress concentration layers we use stretched coordinates x and y/ε, or x/c and y, in layers along the x and y directions, respectively. In terms of complex variables, these pairs of coordinates are

$$Z_1 = z_1/c \quad\text{and}\quad Z_2 = z_2/\varepsilon. \tag{3.6}$$

For the stress concentration layer along the line y = 0, the theory of stress concentration layers ([1], Section 10) gives a small stretching displacement u whose form near a crack tip is

$$u = 2A\varepsilon \operatorname{Re}\sqrt{iZ_2}\,. \tag{3.7}$$

This is the same as the displacement given by the elastic potential $K_u \chi_u$, asymptotically as ε approaches zero. Stress concentration layer theory shows how to express the coefficient A in terms of the tip force F_0 given by the inextensible theory ([1], Section 10):

$$A^2 = F_0^2/\pi L^* G^2. \tag{3.8}$$

Here L^* is defined by

$$\frac{1}{L^*} = \frac{1}{2}\left(\frac{1}{L_+} + \frac{1}{L_-}\right), \tag{3.9}$$

in the notation of Figure 1. With this identification of A, the tensile stress σ_{xx} is then given to leading order in ε, on the line ahead of the crack, by

$$\sigma_{xx} = F_0/(\pi\varepsilon L^* r)^{1/2}. \tag{3.10}$$

The coefficient multiplying $(2\pi r)^{-1/2}$ is the *stress intensity factor* as usually defined. F_0 is the total tensile force in the layer passing in front of the tip of the crack, and εL^* is an order of magnitude estimate of the thickness of the layer.

In the vertical layer along the line $x = 0$, stress concentration layer theory gives a small stretching displacement v whose form near the crack tip is

$$v = -2Bc \operatorname{Re}(i\sqrt{Z_1})\ , \tag{3.11}$$

which is the same as that given by the elastic potential $K_v \chi_v$ for small c. By a J-integral method like that used to determine A, it is found that

$$B^2 = N_0^2/\pi H^* G^2\ , \tag{3.12}$$

where

$$\frac{1}{H^*} = \frac{1}{2}\left(\frac{1}{H_+} + \frac{1}{H_-}\right). \tag{3.13}$$

The corresponding tensile stress σ_{yy} is then

$$N_o/\sqrt{\pi c H^* r} \tag{3.14}$$

on the line ahead of the tip.

By using stress concentration layer theory as a bridge between elasticity theory and the inextensible theory, we have found that the elastic stress intensity factors can be determined, to lowest order in ε and c, when $F_o/\sqrt{L^*}$ and $N_o/\sqrt{H^*}$ are known (and θ_o is given). The relation is not quite complete because we assumed that $\sin\theta_o$ and $\cos\theta_o$ are not zero. When $\sin\theta_o = 0$, $K_u\chi_u$ gives a pure sliding mode which is not possible at all in the inextensible theory, and when $\cos\theta_o = 0$, $K_v\chi_v$ is a pure sliding mode. These cases are presumably of no importance. With those exceptions, knowledge of $F_o/\sqrt{L^*}$ and $N_o/\sqrt{H^*}$ is equivalent to knowledge of the elastic stress intensity factors.

4. Fracture criteria.

A *fracture criterion* is a constitutive equation that states the conditions under which a crack can become longer, and the direction of its advance when those conditions are satisfied. The fracture criterion for a given material is a property of that material, and like other material properties, fracture criteria should be expected to take different forms in different materials.

The Griffith criterion is based on the fact that the mechanical energy released when a crack opens must be sufficient to provide the surface energy (surface tension) for the fresh surface that appears in the process. As it stands, this statement is not a fracture criterion, although it implies the negative result that there can be no fracture if the incipient energy release rate is not large enough. The Griffith criterion states that if the mechanical energy release rate is large enough to provide the surface energy, then fracture will indeed take place. Under this assumption, the direction of crack advance is that for which the incipient energy release rate is largest, since that is the direction for which the fracture criterion will first be satisfied.

The Griffith criterion may accurately describe the behavior of some brittle materials, but it is obviously irrelevant in ductile fracture of metals and tearing of fabrics. Like any other constitutive equation, it may be valid for some materials, in some circumstances, but it is not a universal law of physics.

In fabrics, tears progress by breaking threads. It is natural to postulate that there is a critical breaking force, and that a thread will break when the tension in it reaches that critical value. The direction of propagation of the tear is, by definition, the direction perpendicular to the thread that broke. In principle, threads in the warp and woof might break simultaneously, so that the tear would cut across both families of fibers, but I have never seen such a thing. Tears usually run parallel to one family of fibers, breaking the fibers in the other family. The direction of the tear may change suddenly, so that again only one family of fibers is being broken after the turn. Stiff plates made of cross-ply laminates show the same sort of behavior.

Within the idealized theory considered here, in an infinitesimal extension of the crack in the direction θ, beginning from a configuration like that in Figure 1, the mechanical energy loss per unit length of crack extension is [3]

$$\mathcal{G} = (F_0^2/GL^*)(\sin\theta)_+ + (N_0^2/GH^*)(\cos\theta)_+ \, . \qquad (4.1)$$

Here f_+ means the positive part of f, equal to f when f is positive and equal to zero when f is negative. Thus, for example, the force F_0 in the fiber $y = y_0$ contributes to the energy release rate when the crack extension cuts that fiber, and only then.

From (4.1), evidently every fracture criterion will lead to a non-negative energy release rate, because $\mathcal{G}$ is intrinsically non-negative. Thus the requirement of a non-negative energy release rate imposes no restriction on the fracture criterion.

A proper fracture criterion must lead to crack opening, rather than interpenetration of the parts of the body on opposite sides of the crack. Near the tip, the distance between opposite sides of the crack is proportional to distance from the tip (in the present theory), with the proportionality factor [3]

$$(2F_0/GL^*)\sin^2\theta + (2N_0/GH^*)\cos^2\theta . \qquad (4.2)$$

The requirement that the opening rate must be non-negative gives a mild restriction on the fracture criterion. For example, under the assumption that fracture occurs when $\mathcal{G}$ reaches a critical value, fracture could occur with both F_0 and N_0 negative; the requirement of a positive opening rate rules this out.

Fracture criteria will involve functions of those parameters that are sufficient to specify the state of stress near the tip of the crack and that are variable, in the sense that they do not have fixed values for a given material. Constants like G, ε, and c, which specify a particular material within a given class of materials, may be involved in the fracture criterion, but they are not to be treated as variables.

For example, in what we call the *critical force* criterion, the fiber $y = y_0$ breaks when F_0 reaches a critical value F_c, and the crack extends in the y-direction; the fiber $x = x_0$ breaks when N_0 reaches a critical value N_c, and the crack extends in the x-direction. The variables, used to describe the stress state near the crack tip, are F_0 and N_0. The parameters F_c and N_c select a particular material from the class of those that satisfy the critical force criterion.

One general way to specify a fracture criterion is to define a function $P(F_0,...;\theta)$ in which the first set of variables describes the stress state near the tip, and θ is the direction of incipient crack advance. Then we can specify both the conditions for crack propagation and the direction of crack advance by saying that the crack will advance in the direction θ that maximizes $P(...;\theta)$ when that maximum reaches some critical value P_c.

For example, the energy release rate (4.1) can be used as a fracture potential P. The variables, describing the stress state near the tip, are F_0^2/L^* and N_0^2/H^*. The present direction of the crack, θ_0, is a variable that we would expect to see in a fracture criterion, but it does not appear in this one. The angle θ is not the present crack direction, but the direction of incipient advance. If conditions are such that the requirement of a positive opening rate (4.2) imposes no restriction on θ , the direction θ that maximizes $\mathcal{G}$ is given by

$$\tan\theta = F_0^2 H^*/N_0^2 L^* , \tag{4.3}$$

and the energy release rate for that direction is

$$\mathcal{G} = G^{-1}[(F_0^2/L^*)^2 + (N_0^2/H^*)^2]^{1/2}. \tag{4.4}$$

The *energy release rate* criterion states that the crack will advance in the direction given by (4.3) when (4.4) reaches a critical value $\mathcal{G}_c$.

We have emphasized the interpretation of the idealized theory as an asymptotic approximation to elasticity theory. It should be borne in mind that the idealized theory may also be applicable to materials for which classical elasticity is not a good approximation. However, let us continue to consider cases in which we can regard elasticity theory as a more accurate approximation. In such cases, the stress components become infinite at the tip of a crack, and from Section 3 we find that the variables used in specifying the stress singularity are $F_0/\sqrt{L^*}$, $N_0/\sqrt{H^*}$, and θ_0, the present direction of the crack. Then a fracture potential should be a function of the form

$$P(F_0/\sqrt{L^*},\ \ N_0/\sqrt{H^*},\ \ \theta_0;\ \theta). \tag{4.5}$$

For example, the energy release rate is a function of this form.

A fracture criterion that has seen some use is the *maximum opening displacement* criterion, in which the opening rate defined in (4.2) is used as a fracture potential. But the geometrical factors L^* and H^* do not enter into the definition of the opening rate in the form shown in (4.5), so under this criterion, the conditions for fracture depend not only on the stress state at the tip of the crack but also on the location of the crack tip within the body. The same stress state may or may not produce fracture, depending on conditions other than the stress state at the tip. Thus for elastic fracture analysis, we must reject this criterion.

The critical force criterion was not stated in terms of a fracture potential, although it could be, but it is easy to see that this criterion is not compatible with elasticity theory, for the same reason that the opening displacement criterion is not. The intensity of the stress singularity depends on $F_0/\sqrt{L^*}$, not F_0 alone. This suggests that in cases in which the critical force criterion seems intuitively reasonable, the actual criterion might be what we will call the *critical stress* criterion, that fracture occurs when

$$F_0/\sqrt{L^*} = K_1 \quad \text{or} \quad N_0/\sqrt{H^*} = K_2, \tag{4.6}$$

with crack advance in the y-direction in the first case, and in the x-direction in the second.

All of the fracture criteria that have been mentioned are merely conjectural. The actual fracture criterion for a given material must be determined by experimentation on that material, with the aid of theoretical analysis of the results of the experiments.

5. Crack trajectories [3].

Within a theory in which problems can actually be solved, we can try out various fracture criteria to see what kinds of fractures they predict. As an example for which the analysis is particularly simple, let us consider a rectangular sheet loaded by uniform tensile forces on its ends $x = 0$ and $x = L$,

$$\sigma_{xx}(0,y) = \sigma_{xx}(L,y) = T, \tag{5.1}$$

with no tractions on the edges $y = 0$ and $y = H$. A crack or tear runs from the point $(x^*,0)$ on the lower edge to an interior point (x_0,y_0).

The method of shearing stress resultants shows that for $y > y_0$,

$$u'(y)L + v(L) - v(0) = 0, \tag{5.2}$$

so $u'(y)$ is constant in that region. Then $\sigma_{xy,y} = 0$, so

$$\sigma_{xx,x} = -\sigma_{xy,y} = 0, \tag{5.3}$$

and thus $\sigma_{xx} = T$ for $y > y_0$, except in the singular fiber $y = H$ along the top edge. Assuming that the crack slants toward the right, so that $x_0 \geq x^*$, it can be shown by a similar argument that $\sigma_{yy} = 0$ in the regions $x > x_0$ and $x < x^*$, except in the singular fibers along the edges $x = 0$ and $x = L$.

This information about the stress field is enough to allow us to compute the tip forces F_0 and N_0. Consider the forces that act on the region to the right of the crack and, for $y > y_0$, the line $x = x_0$. By taking moments of these forces around the point (x_0,H) we obtain

$$(H-y_0)F_0 = (H - \tfrac{1}{2} y_0)Ty_0. \tag{5.4}$$

A similar argument gives

$$(L-x_0)N_+ = (y_0/2)Ty_0, \tag{5.5}$$

where we use the notation N_+ for N_0 when the crack slants toward the right. If it slants toward the left so that $x_0 < x^*$, the tip force N_0 is denoted by N_-, and we find that

$$x_0 N_- = (y_0/2)Ty_0. \tag{5.6}$$

If the crack is vertical, along the line $x_0 = x^*$, there are singular fibers on both faces of the crack, and the forces in these fibers at the tip of the crack are N_+ in the right-hand fiber and N_- on the left. Just ahead of the tip, these two fibers coalesce into one, in which the force is $N_+ + N_-$.

The present problem is unusually simple because we can determine the tip forces without knowing the shape of the crack. To determine the shape, we suppose that the crack begins with an arbitrarily short vertical segment along the line $x = x^*$. Then the tension T is varied in such a way as to keep the fracture criterion satisfied, and we suppose that this is done so carefully that there are no dynamic effects. The stress concentration at the tip of the crack is generally more severe, the longer the crack is (for fixed T), so for quasi-static propagation it is necessary to decrease T as the crack grows. However, we do not need to know the loading history in order to compute the trajectory.

Let us first consider the critical force criterion. As we have pointed out, this is not compatible with elasticity theory, but it keeps the algebra simple. Under this criterion, the crack tip moves only in the y-direction or the positive or negative x-direction, and the choice depends on which is larger, F_o/F_c or N_o/N_c, since we suppose that T is kept large enough to make one these ratios equal to unity. From (5.4) and (5.6), F_o/F_c is equal to N_-/N_c on the curve

$$x = (F_c/N_c)y(H-y)/(2H-y). \tag{5.7}$$

This curve is sketched in Figure 2. To the left of that curve, N_-/N_c is greater than F_o/F_c, so trajectories in that region run in the negative x-direction. Similarly, N_+/N_c is greater than F_o/F_c in the region to the right of the curve

$$x = L - (F_c/N_c)y(H-y)/(2H-y), \tag{5.8}$$

so cracks in that region run in the positive x-direction. In the region between these two curves, cracks go in the y-direction.

A crack starting from the bottom edge of the sheet propagates in the y-direction until the tip reaches one of the curves (5.7) or (5.8), and then turns abruptly toward the left or right. If the crack along $x = x^*$ does not intersect either of the corner loci (5.7) or (5.8), it continues straight across the sheet to $y = L$. The maximum value of x on (5.7) is

$$x_m = (3-2\sqrt{2})H(F_c/N_c), \quad 3-2\sqrt{2} = 0.17, \tag{5.9}$$

so the crack will be L-shaped with a turn to the left if $x^* < x_m$, L-shaped with a turn to the right if $x^* > L-x_m$ and straight across the sheet if $x_m < x^* < L-x_m$ (assuming that the sheet is long enough that these are distinct regions). When $L/H < 0.34\ (F_c/N_c)$, x_m is greater than L/2 so no crack can run all the way across the sheet. Thus, if

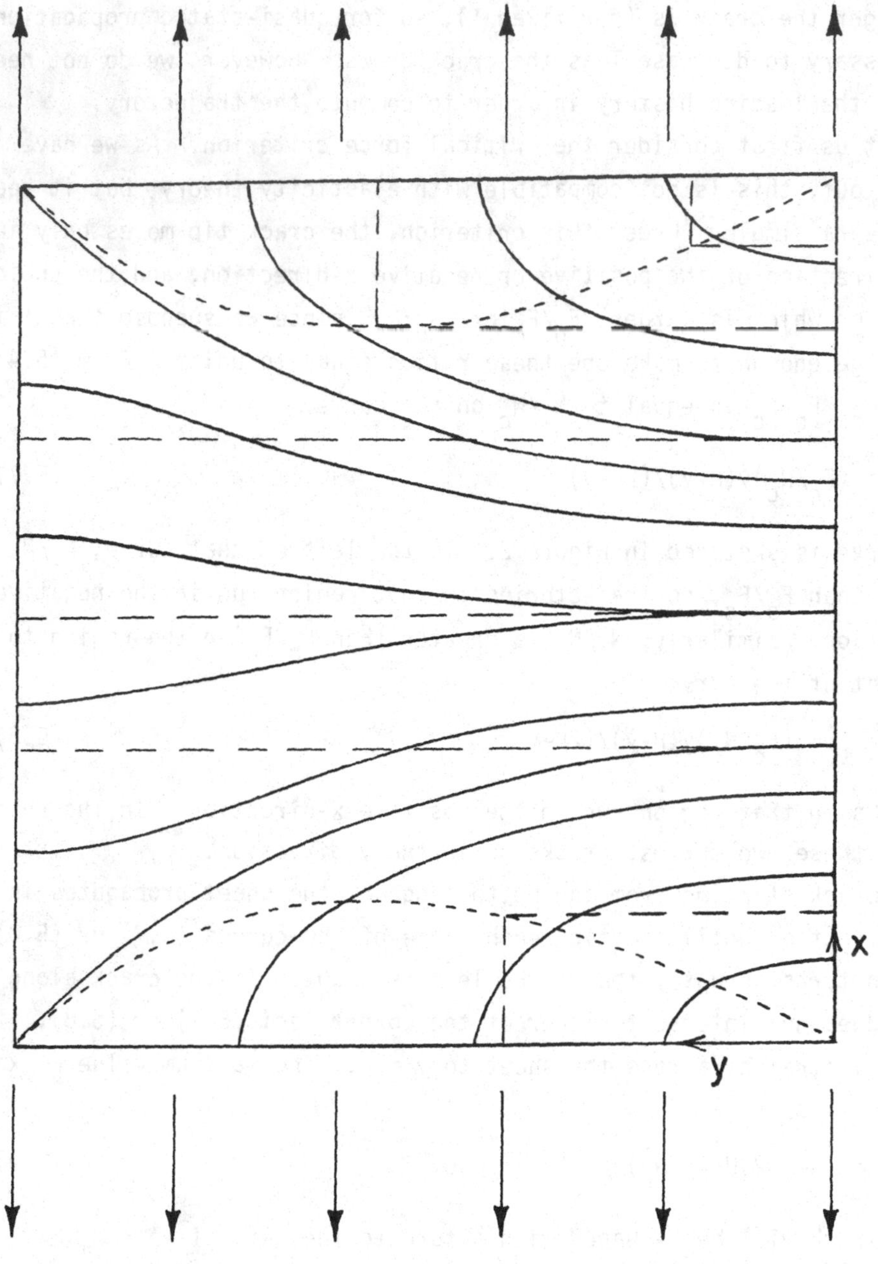

Figure 2

$F_c = N_c$, it is not possible to tear the sheet in half (with boundary loads of the kind we have specified) if H is more than three times L.

Crack trajectories under the critical stress criterion (4.6) are straight or L-shaped, like those obtained with the critical force criterion. The corner loci (5.7) and (5.8) are modified by replacing F_c by $K_1\sqrt{L^*}$ and N_c by $K_2\sqrt{H^*}$. In this problem, the harmonic mean lengths L^* and H^* are

$$L^* = 2x_o(L-x_o)/L \quad \text{and} \quad H^* = 2y_o(H-y_o)/H, \tag{5.10}$$

so the locus (5.8) is replaced by

$$x/(L-x) = Ry(H-y)/(2H-y)^2, \quad R = K_1^2H/K_2^2L, \tag{5.11}$$

which is qualitatively similar to (5.8). The largest value of x on this locus is

$$x_m = \frac{1}{8}\,H(K_1/K_2)^2/[1 + \frac{1}{8}\,(H/L)(K_1/K_2)^2]. \tag{5.12}$$

This is roughtly like (5.9) but depends weakly on L.

The maximum energy release rate criterion predicts smooth crack trajectories that cut across both families of fibers simultaneously. The slope of those that slope to the right is, from (4.3),

$$dy/dx = (F_o/N_o)^2(H^*/L^*)$$
$$= \frac{L}{H}\,\frac{L-x}{x}\,\frac{(2H-y)^2}{y(H-y)}\,. \tag{5.13}$$

Although this differential equation can be integrated explicitly, the nature of the trajectories is easier to see by drawing the direction field. Cracks start vertically at the bottom edge and exit vertically at the top edge, and go nearly straight across if they start far enough from the corner (L,0). These that start near this corner turn to the right and exit horizontally through the edge x = L. The cracks that slope to the left are the mirror images of these. The left-sloping cracks have a larger energy release rate in the region x < L/2. Thus the pattern of possible crack trajectories is generally like that found with the critical stress criterion, but the cracks do not follow the fibers. This pattern of cracks appears plausible for a brittle material, but not for tears in fabrics.

6. Dynamic crack propagation [4].

In dynamic problems the stress and displacement components satisfy the momentum equations,

$$\sigma_{xx,x} + \sigma_{xy,y} = \rho u_{tt},$$
$$\sigma_{yx,x} + \sigma_{yy,y} = \rho v_{tt}, \tag{6.1}$$

where $u = u(y,t)$ and $v = v(x,t)$. With the shearing stress σ_{xy} given by (2.4), the momentum equations can be integrated formally to find the reactions σ_{xx} and σ_{yy}:

$$\sigma_{xx} = x(\rho u_{tt} - Gu_{yy}) + f(y,t),$$
$$\sigma_{yy} = y(\rho v_{tt} - Gv_{xx}) + g(x,t). \tag{6.2}$$

Let $x_0(t)$ and $y_0(t)$ be the coordinates of the tip of a growing crack. The components of velocity of the tip are

$$U = x_0'(t), \quad V = y_0'(t). \tag{6.3}$$

The particle velocity u_t and amount of shear u_y will generally be discontinuous across the fiber $y = y_0(t)$ through the tip, but u is continuous, so $u_0(y_0(t),t) = u_\pm(y_0(t),t)$, in the notation of Figure 1. Differentiating this continuity condition with respect to time gives

$$\Delta u_t + V\Delta u_y = 0, \tag{6.4}$$

where Δu_t and Δu_y are the jumps in u_t and u_y. Similarly,

$$\Delta v_t + U\Delta v_x = 0. \tag{6.5}$$

The discontinuous parts of u_t and u_y can be written as $(\Delta u_t)H(y-y_o(t))$ and $(\Delta u_x)H(y-y_o(t))$, where H is the unit step function. By using these expressions in (6.2) we find that, as in the static case, the tensile stress σ_{xx} is singular in the fiber $y = y_o(t)$,

$$\sigma_{xx} = F(x,t)\delta(y-y_o(t)) + \ldots \, . \tag{6.6}$$

The force F satisfies

$$F_{,x} = -\rho V\Delta u_t - G\Delta u_y \, . \tag{6.7}$$

Similarly, there is a finite force N(y,t) in the fiber $x = x_o(t)$, which satisfies

$$N_{,y} = -\rho U\Delta v_t - G\Delta v_x \, . \tag{6.8}$$

For a crack situated like that in Figure 1, suppose that finite traction is prescribed at the point (x_o-L_-,y_o), so that F = 0 there. Then the value of F at the tip of the crack is

$$F_o = -L_-[\rho V\Delta u_t + G\Delta u_y]. \tag{6.9}$$

Similarly, with N = 0 at (x_o,y_o+H_+), the value of N at the crack tip is

$$N_o = H_+[\rho U\Delta v_t + G\Delta v_x]. \tag{6.10}$$

By eliminating the velocity jumps from (6.9) and (6.10) by using (6.4) and (6.5), we obtain

$$F_o = -GL_-[1 - (V/c)^2]\Delta u_y \tag{6.11}$$

and

$$N_o = GH_+[1 - (U/c)^2]\Delta v_x. \tag{6.12}$$

Here c is the speed of shear waves, defined by

$$c^2 = G/\rho. \tag{6.13}$$

7. Quasi-static and steady propagation of a straight crack.

Earlier we considered a crack trajectory problem that was simple because the tip forces could be determined before the crack trajectory was known. We now consider a problem that is simple in a different way; the trajectory can be guessed in advance, and only the speed of propagation is unknown.

We consider a crack that begins to propagate from the point (0,0) on the boundary of the quarter-space $x \geq 0$, $y \leq H$. The crack is opened by a displacement $v(0,t) = D(t)$ on the part of the boundary $x = 0$ for which $y \geq 0$. There is no normal traction on this end, and no traction on any other part of the boundary. We will use the critical force or critical stress fracture criterion.

It is an easy and correct assumption that the crack will propagate straight along the line $y = 0$, at least at first. So long as this is so, the displacement component u is zero everywhere, and v is zero except in the region above the crack, $y \geq 0$, $0 \leq x \leq x_o(t)$. The coordinate $x_o(t)$ of the tip is unknown.

When the opening displacement D(t) increases sufficiently gradually, we can treat the problem quasi-statically. In the static solution with opening displacement D and the crack tip at x_o, the deformed region is in a state of simple shear with $v_x = -D/x_o$. Then the tip force N_o, found by neglecting $(U/c)^2$ in (6.12), is $N_o = GHD/x_o$. Let us use the critical force criterion and define

$$M = N_c/GH. \tag{7.1}$$

Then so long as the amount of shear D/x_o is less than M, crack propagation does not take place. If D(t) increases steadily, the tip $x_o(t)$ must move forward since otherwise D/x_o would exceed M. Thus

$$x_o(t) = D(t)/M. \tag{7.2}$$

More generally, if $D^*(t)$ is the largest opening displacement that has occurred up to time t, $x_o = D^*/M$.

If the tip velocity U calculated from (7.2) is not so small that $(U/c)^2$ is negligible, the previous analysis must be modified. To consider possible inertial effects on v, we look at the (integrated) momentum equation (6.2b). With $\sigma_{yy} = 0$ on $y = 0$ and $y = H$, this equation implies that v satisfies the wave equation,

$$v_{tt} = c^2 v_{xx} . \tag{7.3}$$

We wish to obtain a solution in the region $0 \le x \le x_0(t)$, satisfying

$$v(x_0,t) = 0 \text{ and } [1 - (U/c)^2]v_x(x_0,t) = -M \tag{7.4}$$

at the tip, and

$$v(0,t) = D(t) \tag{7.5}$$

at $x = 0$. The general solution satisfying (7.4) can be expressed in terms of a function of one variable, f, and (7.5) then leads to a functional equation for f [5].

To deal with an exact solution, suppose that the crack opens with a constant speed s: $D = st$. Then the equation and two of the boundary conditions are satisfied by

$$v(x,t) = s(t-x/U), \quad x_0(t) = Ut, \tag{7.6}$$

and the constant speed of propagation U is determined by the fracture criterion (7.4b):

$$\begin{aligned} M &= [1 - (U/c)^2](s/U) \\ &= (s/U) - (U/s)(s/c)^2. \end{aligned} \tag{7.7}$$

When s/c is small we recover the quasi-static result $U = s/M$.

The tip speed increases as the opening speed increases, but from (7.7) it is evident that the tip speed U cannot exceed the signal speed c. When s/c is large (which would require explosive loading),

$$U/c \cong 1 - (M/2)(c/s). \tag{7.8}$$

8. Quasi-static propagation of a curved crack.

In the problem considered in Section 7, the crack eventually deviates from the straight line $y = 0$. We will attempt to analyze the ensuing motion quasi-statically, but will find that this is not entirely possible.

If D increases slowly enough that inertial effects can be neglected, the amount of shear stays at the maximum static value M, so the shearing stress is GM. Since there is no shearing traction on the crack face, the fiber $y = 0$ is singular, and the force in it at the crack tip is $F_0 = GMx_0(t)$. This eventually reaches the breaking force F_c, at time t_1, say. Then as $D(t)$ continues to increase, the crack trajectory veers upward along a curve $y = y_c(x)$ or $x = x_c(y)$.

In the region $0 \leq x \leq x_c(y)$, $0 < y \leq y_0(t)$, the displacement component u is no longer 0. It is easy to see that all conditions are satisfied (statically) if this region rotates rigidly:

$$u = (D/x_0)(y-y_0), \quad v = (D/x_0)(x_0-x). \tag{8.1}$$

For, the shearing stress in this region is then zero, so the condition of no traction on the crack surface can be satisfied.

After the crack has turned away from the line $y = 0$, the tip force F_0 is given by (6.11). Neglecting the inertial term $(V/c)^2$ here gives $F_0 = GD(t)$. With $GD(t_1) = F_c$, if D increases beyond the value it had when the fiber $y = 0$ broke then F_0 becomes larger than F_c, which is impossible.

This merely means that when $D(t) > D(t_1)$ the problem has no static solution. When there is no static solution, one might expect catastrophic failure, and I would have thought that this would mean that the solution would become fully dynamical. However, this is not necessarily so. It is necessary to retain the inertial term in the expression for F_0, so that the failure condition $F_0 = F_c$ becomes

$$F_c/G = D(t)[1 - (V/c)^2], \tag{8.2}$$

but the expression for N_0 can still be treated quasi-statically:

$$N_0 = GH(1 - y_0/H)D/x_0. \qquad (8.3)$$

To see that this is possible, suppose that D is increased just a little beyond the value F_c/G required to break the fiber $y = 0$, and then held fixed. From (8.2), $y_0(t)$ increases at the rate

$$y_0'(t)/c = [1 - (F_c/GD)]^{1/2}. \qquad (8.4)$$

Because of the factor c, y_0' will be large even if D is only moderately larger than F_c/G, but y_0' can theoretically be made arbitrarily small by letting D exceed F_c/G by a sufficiently small amount. In the latter case, $y_0(t)$ grows slowly but steadily, so from (8.3) we see that N_0 drops below the value $GHD(t_1)/x_0(t_1) = N_c$, so x_0 cannot increase further.

Even though the crack turns through a right angle, it does not actually have a sharp corner. During the short time interval while D is increasing from $D(t_1)$ to its slightly larger final value, the value of $x_0(t)$ found by setting $N_0 = N_c$ in (8.3) is

$$x_0(t) = D(t)[1 - y_0(t)/H]/M. \qquad (8.5)$$

From (8.4) it follows that y_0 increases more slowly than D does, so the right-hand member of (8.5) increases a little before it begins to decrease. During the interval while it is increasing, the crack cuts across both families of fibers simultaneously.

ACKNOWLEDGMENT

This paper was prepared with the support of a grant MCS 79-03392 from the National Science Foundation. We gratefully acknowledge this support.

REFERENCES

[1] PIPKIN, A.C. Stress channelling and boundary layers in strongly anisotropic solids. This volume, Chapter IV.

[2] PIPKIN, A.C. Stress analysis for fiber-reinforced materials. Adv. in Appl. Mech. 19, (1979) 1-51.

[3] PIPKIN, A.C. and ROGERS, T.G. Crack paths in sheets reinforced with two families of inextensible fibers. Mech. Today 5, (1980) 397-407.

[4] MANNION, L.F. Dynamic fracture of idealized fiber-reinforced materials. Ph.D. Thesis, Division of Applied Mathematics, Brown University, 1981. J. Elast., Forthcoming.

[5] PIPKIN, A.C. Crack speeds in an ideal fiber-reinforced material. Q. Appl. Math. 42 (1984).

VII

REINFORCEMENT OF HOLES IN PLATES BY FIBRE-REINFORCED DISCS

A. J. M. SPENCER

Department of Theoretical Mechanics
University of Nottingham
Nottingham, NG7 2RD
England

1. INTRODUCTION

A potentially useful method of reinforcing plates containing holes is by bonding annular discs of cylindrically wound fibre-reinforced material to the lateral surfaces of the plate, as illustrated in Figure 1. Reinforcement of this kind offers considerable potential weight-saving advantages in aircraft and other structures.

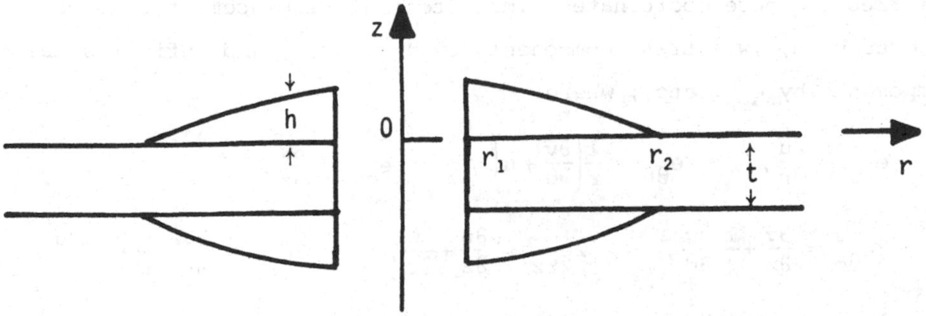

Figure 1. Reinforcement of a hole in a plate by annular discs

For optimal design of the reinforcement an analysis of the stress and deformation in the plate and reinforcement is clearly necessary. McKenzie and Webber [1] gave a two-dimensional analysis of the following two problems for a circular reinforced hole:

Problem I Uniform all-round tension at infinity,

Problem II Pure shear at infinity.

Appropriate superposition of the solutions to these two problems gives the solution for any uniform stress state at infinity, and in particular for the problem of uniaxial tension at infinity. Mansfield [2] developed a three-dimensional analysis of Problem I, for the case of uniform reinforcement thickness, but his method requires extensive computation to yield numerical results. Subsequently Mansfield [3] gave an approximate method of solution of Problems I and II, also for uniform reinforcement thickness; this method in effect reduces the three-dimensional problem to an equivalent two-dimensional problem.

We here present an approximate three-dimensional analysis of Problems I and II which differs somewhat from that of Mansfield [3], although the methods are related. The approximation is based on the assumption that the reinforcement is made of a fibre-reinforced material in which the ratios of the axial and transverse shear moduli to the fibre extension modulus can be regarded as small parameters.

1.1 Formulation

We employ cylindrical polar coordinates r,θ,z (see Figure 1). Referred to these coordinates, infinitesimal displacement components are denoted by u,v,w, stress components by σ_{rr}, etc., and infinitesimal strain components by e_{rr}, etc., where

$$\begin{aligned} e_{rr} &= \frac{\partial u}{\partial r}, \qquad e_{\theta\theta} = \frac{1}{r}\left(\frac{\partial v}{\partial \theta} + u\right), \qquad e_{zz} = \frac{\partial w}{\partial z}, \\ 2e_{\theta z} &= \frac{\partial v}{\partial z} + \frac{1}{r}\frac{\partial w}{\partial \theta}, \qquad 2e_{rz} = \frac{\partial u}{\partial z} + \frac{\partial w}{\partial r}, \qquad 2e_{r\theta} = \frac{\partial v}{\partial r} - \frac{v}{r} + \frac{1}{r}\frac{\partial u}{\partial \theta}. \end{aligned} \tag{1}$$

The plate has thickness t and contains a circular hole of radius r_1, so that the plate occupies the region $-t < z < 0$, $r > r_1$. The reinforcement is symmetrically placed on either side of the plate and is rigidly bonded to the plate at the surfaces $z = 0$, $z = -t$ over the annulus $r_1 < r < r_2$. The thickness of the reinforcement is not necessarily uniform, and is denoted by $h(r)$. In practical problems we expect to have $h/r_1 \ll 1$, and $t/r_1 \ll 1$. By symmetry, it is sufficient to consider the reinforcement attached to the upper face of the plate.

It is assumed that the surface $r = r_1$ of the hole is free from traction:

$$\sigma_{rr} = 0, \quad \sigma_{r\theta} = 0, \quad \sigma_{rz} = 0, \quad \text{on} \quad r = r_1, \tag{2}$$

and that the surface $r = r_2$ of the reinforcement is traction-free:

$$\sigma_{rr} = 0, \quad \sigma_{r\theta} = 0, \quad \sigma_{rz} = 0, \quad r = r_2, \quad 0 < z < h(r_2). \tag{3}$$

The lateral surfaces of the plate and reinforcement are also supposed to be free of traction, so that

$$\sigma_{zz} = 0, \quad \sigma_{rz} = 0, \quad \sigma_{\theta z} = 0 \quad \text{for} \quad \begin{cases} z = h(r), & r_1 < r < r_2, \quad (4) \\ z = 0, & r > r_2. \quad (5) \end{cases}$$

For (4) to hold, it is assumed that the slope of the reinforcement surface $z = h(r)$ is small, i.e. that $h'(r) \ll 1$.

We consider two basic problems:

Problem I. All-round tension P at infinity:

$$\sigma_{rr} \to P, \quad \sigma_{\theta\theta} \to P, \quad \sigma_{r\theta} \to 0 \quad \text{as} \quad r \to \infty. \tag{6}$$

Problem II. Pure shear Q at infinity:

$$\sigma_{rr} \to Q\cos 2\theta, \quad \sigma_{\theta\theta} \to -Q\cos 2\theta, \quad \sigma_{r\theta} \to -Q\sin 2\theta \quad \text{as} \quad r \to \infty. \tag{7}$$

Then, for example, the case of uniaxial tension at infinity, of magnitude X acting in the direction $\theta = 0$, is solved by setting $P = \frac{1}{2}X$, $Q = \frac{1}{2}X$, and superposing the solutions to Problems I and II.

At the interfaces between the plate and the reinforcement, we require continuity of displacement and of σ_{zz}, σ_{rz}, $\sigma_{\theta z}$.

In view of the boundary conditions (4) and (5), and the assumed thinness of the plate and reinforcement, we assume that σ_{zz} is negligible throughout the plate and reinforcement. It is then not possible to satisfy exactly the equation for equilibrium in the z-direction, but the approximation involved is justifiable. Because shear-lag effects may be significant, we make no assumption regarding σ_{rz} and $\sigma_{\theta z}$ at this stage.

The plate is assumed to be of isotropic elastic material, with Lamé constants λ and μ. Then, with $\sigma_{zz} = 0$, the relevant stress-strain relations are

$$
\begin{aligned}
(\lambda+2\mu)\sigma_{rr} &= 4\mu(\lambda+\mu)e_{rr} + 2\lambda\mu e_{\theta\theta}\,, \\
(\lambda+2\mu)\sigma_{\theta\theta} &= 2\lambda\mu e_{rr} + 4\mu(\lambda+\mu)e_{\theta\theta}\,, \\
\sigma_{\theta z} = 2\mu e_{\theta z}\,, \qquad &\sigma_{rz} = 2\mu e_{rz}\,, \qquad \sigma_{r\theta} = 2\mu e_{r\theta}\,.
\end{aligned}
\tag{8}
$$

The reinforcement is regarded as consisting of cylindrically wound stiff fibres in a relatively low-modulus matrix. The fibre lie in circles r = constant in the planes z = constant. The composite is assumed to be transversely isotropic with respect to the fibre (θ) direction. With $\sigma_{zz} = 0$, the stress-strain relation in the reinforcement takes the form

$$
\begin{pmatrix} \sigma_{\theta\theta} \\ \sigma_{rr} \\ \sigma_{rz} \\ \sigma_{\theta z} \\ \sigma_{r\theta} \end{pmatrix}
=
\begin{pmatrix}
c'_{11} & c'_{12} & 0 & 0 & 0 \\
c'_{12} & c'_{22} & 0 & 0 & 0 \\
0 & 0 & 2\mu_T & 0 & 0 \\
0 & 0 & 0 & 2\mu_L & 0 \\
0 & 0 & 0 & 0 & 2\mu_L
\end{pmatrix}
\begin{pmatrix} e_{\theta\theta} \\ e_{rr} \\ e_{rz} \\ e_{\theta z} \\ e_{r\theta} \end{pmatrix},
\tag{9}
$$

where μ_T and μ_L are transverse and longitudinal shear moduli, and c'_{11}, c'_{22}, c'_{12} are elastic constants which can be related to the stiffness components $c_{ijk\ell}$. For the values given in equation (15) of Chapter I for a typical carbon fibre - epoxy resin composite we have, in units of $10^9 Nm^{-2}$,

$$
c'_{11} = 239.91\,, \quad c'_{12} = 2.04\,, \quad c'_{22} = 7.56\,, \quad \mu_L = 5.66\,, \quad \mu_T = 2.46\,.
\tag{10}
$$

We note that c'_{11} is nearly equal to the fibre extension modulus E_L, and henceforth we adopt the approximation

$$c'_{11} = E_L \, . \tag{11}$$

The remaining equilibrium equations expressed in cylindrical polar coordinates are

$$\frac{\partial\sigma_{rr}}{\partial r} + \frac{1}{r}\frac{\partial\sigma_{r\theta}}{\partial\theta} + \frac{\partial\sigma_{rz}}{\partial z} + \frac{\sigma_{rr}-\sigma_{\theta\theta}}{r} = 0 \, , \tag{12}$$

$$\frac{\partial\sigma_{r\theta}}{\partial r} + \frac{1}{r}\frac{\partial\sigma_{\theta\theta}}{\partial\theta} + \frac{\partial\sigma_{\theta z}}{\partial z} + \frac{2\sigma_{r\theta}}{r} = 0 \, . \tag{13}$$

Mansfield [2, 3] adopts a slightly different procedure in that he assumes $w = 0$. It is then not necessary to neglect the equilibrium equation in the z-direction, but the boundary condition $\sigma_{zz} = 0$ on the lateral surfaces is not satisfied. Mansfield corrects for this by modifying the elastic constants, and thus obtains equations which are identical to those given above.

2. UNIFORM ALL-ROUND TENSION AT INFINITY

We now consider Problem I, with boundary conditions (6). The problem has axial symmetry, so that $v = 0$, $\sigma_{r\theta} = 0$, $\sigma_{\theta z} = 0$, all variables are independent of θ, and the equilibrium equation (13) is identically satisfied.

2.1 Stress and deformation in the reinforcement

By setting $v = 0$ and substituting (1) and (9) in the equilibrium equation (12), we obtain

$$c'_{22}\left(\frac{\partial^2 u}{\partial r^2} + \frac{1}{r}\frac{\partial u}{\partial r}\right) - E_L\frac{u}{r^2} + \mu_T\left(\frac{\partial^2 u}{\partial z^2} + \frac{\partial^2 w}{\partial r\partial z}\right) = 0 \, .$$

We now assume that w does not vary rapidly with r, and neglect the term $\partial^2 w/\partial r\partial z$, so that the equation becomes

$$c'_{22}\left(\frac{\partial^2 u}{\partial r^2}+\frac{1}{r}\frac{\partial u}{\partial r}\right)-E_L\frac{u}{r^2}+\mu_T\frac{\partial^2 u}{\partial z^2}=0\,. \tag{14}$$

Mansfield [2] solved (14) by separation of variables. This leads to solutions in terms of series of Bessel functions whose orders are, in general, not integers. The computation required to obtain numerical values is therefore quite extensive. We seek a simpler method of obtaining approximate solutions of (14). We note that for many composites c'_{22}/E_L is small (for example, $c'_{22}/E_L \simeq 0.04$ for the composite whose properties are given by equation (15) of Chapter I). Also, we do not expect u to vary rapidly with r, except possibly near $r = r_1$ and $r = r_2$. Therefore, as a first approximation, we neglect the term in c'_{22} in comparison to that in E_L in (14), and approximate (14) by

$$\frac{\partial^2 u}{\partial z^2}-\alpha^2\frac{u}{r^2}=0\,, \tag{15}$$

where

$$\alpha^2 = E_L/\mu_T\,. \tag{16}$$

Although $E_L \gg \mu_T$, it is not permissible to neglect the term in μ_T in (14) because u may vary rapidly with z.

The as yet unknown value of u(r,z) at the interface z = 0 is denoted by U(r). Then the solution of (15) subject to $\sigma_{rz} = 0$ at $z = h(r)$ (and again neglecting $\partial w/\partial r$ compared to $\partial u/\partial z$) is

$$u = U(r)\,\frac{\cosh[\alpha\{h(r)-z\}/r]}{\cosh\{\alpha h(r)/r\}}\,, \tag{17}$$

which gives, in this approximation,

$$\sigma_{\theta\theta} = \frac{E_L U(r)\cosh[\alpha\{h(r)-z\}/r]}{r\cosh\{\alpha h(r)/r\}}\,,$$
$$\sigma_{rz} = -\,\frac{\alpha\mu_T U(r)\sinh[\alpha\{h(r)-z\}/r]}{r\cosh\{\alpha h(r)/r\}}\,. \tag{18}$$

Both $\sigma_{\theta\theta}$ and σ_{rz} take their maximum values at the interface z = 0. These maximum values are

$$\sigma_{\theta\theta} = E_L r^{-1} U(r) \quad \text{at} \quad z = 0,$$
$$\sigma_{rz} = -\alpha\mu_T r^{-1} U(r) \tanh\{\alpha h(r)/r\} \quad \text{at} \quad z = 0. \tag{19}$$

The approximate equation (15) has a simple physical interpretation. Except for terms which are small compared to $\sigma_{\theta\theta}/r$, it is equivalent to

$$\frac{\partial\sigma_{rz}}{\partial z} - \frac{\sigma_{\theta\theta}}{r} = 0. \tag{20}$$

Thus in this approximation the shear stress gradient is assumed to be equilibrated by the hoop tension, and the contribution of σ_{rr} to the radial equilibrium is neglected. This is roughly equivalent to assuming that the shear stress gradient is equilibrated by the fibre tension, and the contribution of the matrix is neglected. It should therefore produce overestimates of $\sigma_{\theta\theta}$, which is on the safe side for design purposes.

2.2 Stress and deformation in the plate

For axial symmetry, (1), (8) and (12) give

$$\frac{\partial^2 u}{\partial r^2} + \frac{1}{r}\frac{\partial u}{\partial r} - \frac{u}{r^2} + \frac{\lambda+2\mu}{4(\lambda+\mu)}\frac{\partial^2 u}{\partial z^2} = 0. \tag{21}$$

Let $\bar{u}(r)$ be the average value of $u(r,z)$ in the plate, so that

$$\bar{u}(r) = \frac{1}{t}\int_{-t}^{0} u(r,z)\, dz. \tag{22}$$

Then by integrating (21) between $z = -t$ and $z = 0$, and using (8) (again neglecting $\partial w/\partial r$ compared to $\partial u/\partial z$), we have

$$\frac{d^2\bar{u}}{dr^2} + \frac{1}{r}\frac{d\bar{u}}{dr} - \frac{\bar{u}}{r^2} + \frac{\lambda+2\mu}{2(\lambda+\mu)}\frac{\sigma_{rz}(r,0)}{\mu t} = 0. \tag{23}$$

For $r > r_2$, this reduces by (5) to

$$\frac{d^2\bar{u}}{dr^2} + \frac{1}{r}\frac{d\bar{u}}{dr} - \frac{\bar{u}}{r^2} = 0,$$

whose solution, subject to (6), is

$$\bar{u}(r) = \frac{(\lambda+2\mu)Pr}{2\mu(3\lambda+2\mu)} + \frac{B}{r},$$

where B is constant. Hence, by eliminating B,

$$\frac{d\bar{u}}{dr} + \frac{\bar{u}}{r} = \frac{P(\lambda+2\mu)}{\mu(3\lambda+2\mu)}, \quad \text{for } r > r_2 . \tag{24}$$

We now consider the solution for (23) in $r_1 < r < r_2$. The condition $\sigma_{rr} = 0$ at $r = r_1$ gives, from (8),

$$\frac{d\bar{u}}{dr} + \frac{\lambda}{2(\lambda+\mu)}\frac{\bar{u}}{r} = 0, \qquad r = r_1 . \tag{25}$$

Also u and σ_{rr} are continuous in r, and it follows that $\bar{u}$ and $d\bar{u}/dr$ are continuous. Hence, from (24),

$$\frac{d\bar{u}}{dr} + \frac{\bar{u}}{r} = \frac{P(\lambda+2\mu)}{\mu(3\lambda+2\mu)}, \qquad r = r_2 . \tag{26}$$

As in Mansfield [2] we now neglect the variation of u through the thickness of the plate and replace the mean displacement $\bar{u}(r)$ by the displacement $U(r)$ at the interface $z = 0$. This amounts to neglecting shear lag effects in the plate. It can be shown that this is justified if $(\pi/\alpha)(r_1^2/ht) \gg 1$, and we assume this to be the case.

By replacing $\bar{u}(r)$ by $U(r)$ and inserting the expression (19) for $\sigma_{rz}(r,0)$, we obtain

$$\frac{d^2U}{dr^2} + \frac{1}{r}\frac{dU}{dr} - \left\{\frac{\gamma(r)}{r}\right\}^2 U = 0, \tag{27}$$

to be solved subject to (25) and (26) (with $\bar{u}$ replaced by U), where

$$\{\gamma(r)\}^2 = 1 + \frac{E_L(\lambda+2\mu)\, r \tanh\{\alpha h(r)/r\}}{2\mu(\lambda+\mu)\alpha t}. \tag{28}$$

When $U(r)$ is determined, then $\sigma_{\theta\theta}$ and σ_{rz} are given in the reinforcement by (18) (σ_{rr} is of less interest, but is easily found if needed) and the mean stress in the plate is given by (8).

It is possible to obtain higher order approximations, but we shall not consider these.

2.3 Numerical results

Equation (27), subject to the stated boundary conditions, is readily solved numerically for a specified reinforcement profile $h(r)$. When h is constant, then for realistic values of the elastic constants and the ratios h/r_1, t/r_1 and r_2/r_1, the function $\gamma(r)$ varies little between $r = r_1$ and $r = r_2$, and $\gamma(r)$ may be approximated by its mean value $\bar{\gamma}$. Then (27) can be solved analytically; this provides a useful check on the calculations.

Mansfield [2] obtained values of the stress components by numerical solution of the full equations for several configurations with h constant.

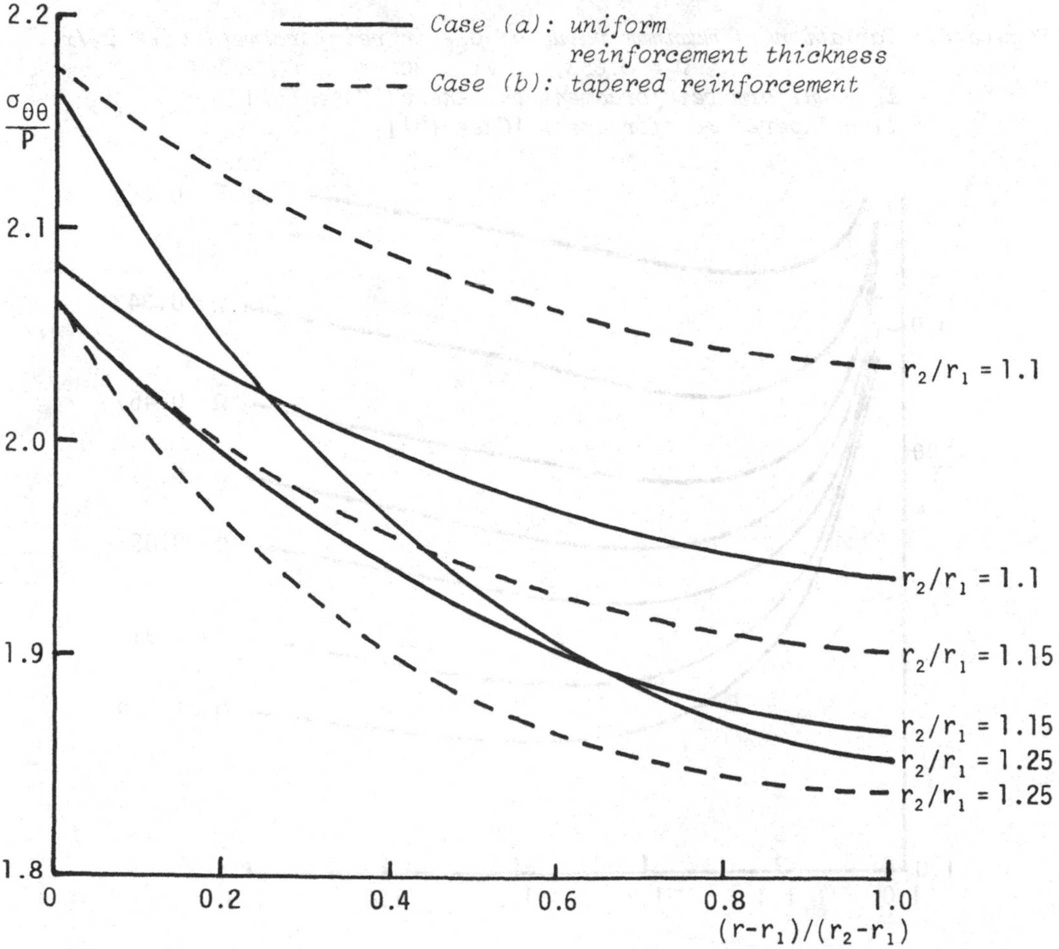

Figure 2. Variation of $\sigma_{\theta\theta}$ at $z = 0$ with distance from the hole.
$\Omega = 0.653$, $r_1/t = 30$

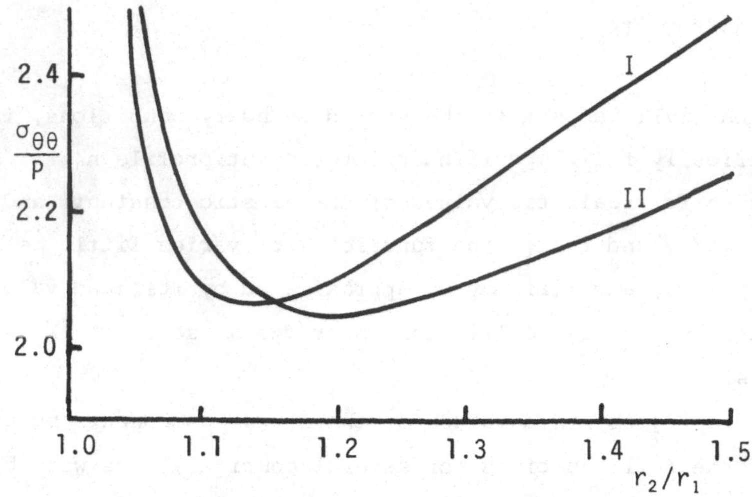

Figure 3. *Variation of maximum value of* $\sigma_{\theta\theta}$ *in reinforcement with* r_2/r_1.
$\Omega = 0.653$, $r_1/t = 30$
I - Uniform reinforcement thickness [Case (a)]
II - Tapered reinforcement [Case (b)]

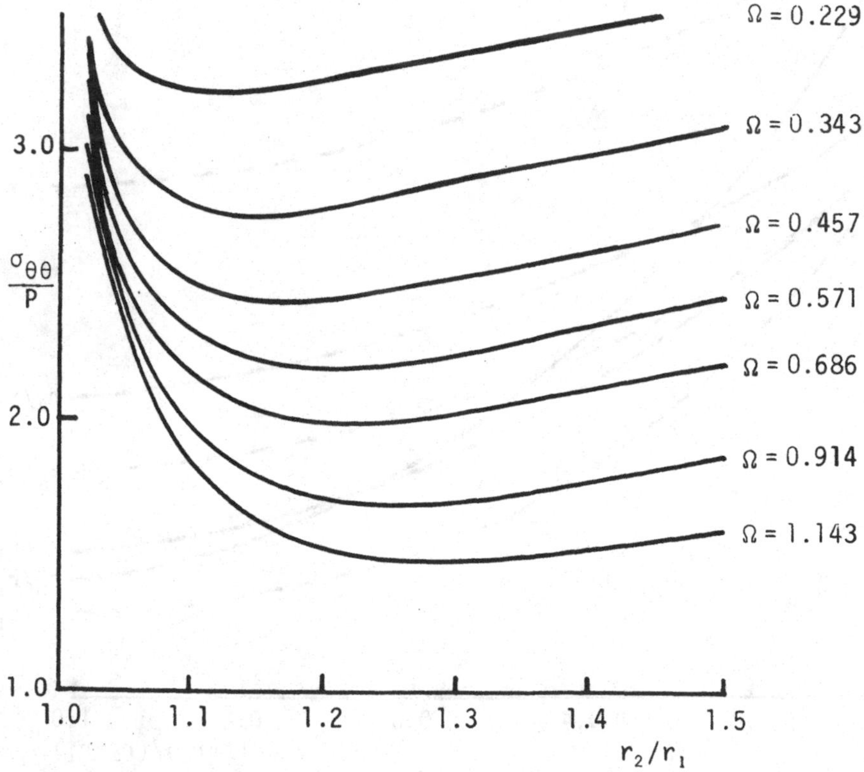

Figure 4. *Variation of maximum value of* $\sigma_{\theta\theta}$ *in reinforcement with* r_2/r_1.
Case (b): tapered reinforcement

The solutions derived from the approximate equation (27) agree with those of [2] to within 1% at the point $(r_1, 0)$ at which $\sigma_{\theta\theta}$ takes its maximum value, to within 2% on the plane $z = 0$, and to within 6% throughout the reinforcement. This agreement is regarded as very satisfactory. Moreover at all points where there is a significant difference between our results and those of Mansfield, the present approximation over-estimates $\sigma_{\theta\theta}$, and so is on the safe side for design purposes.

It is of interest to consider the optimum size and shape of reinforcement. We have considered two cases:

(a) $h(r) = h_0$, where h_0 is constant;

(b) $h(r) = h_1(r_2-r)/(r_2-r_1)$ (i.e. h decreasing linearly from h_1 at $r = r_1$ to zero at $r = r_2$).

Results are expressed in terms of a parameter Ω, where

$$\Omega = \frac{\text{weight of reinforcement}}{\text{weight of material removed from the hole}} .$$

For illustration we use the same numerical values as those in [2], which are (in 10^9Nm^{-2})

$$\lambda = 40.385 , \quad \mu = 26.923 , \quad E_L = 175 , \quad \mu_T = 2.8 , \quad \mu_L = 5.6 ,$$

with (density of plate)/(density of reinforcement) = 1.75 .

Figure 2 shows the variation of $\sigma_{\theta\theta}$ in the reinforcement at $z = 0$ with r. Figure 3 shows how the maximum value of $\sigma_{\theta\theta}$ varies with r_2/r_1, for fixed Ω. It is found that, for given Ω, case (b) gives lower values of the maximum hoop stress than does case (a). For case (b), Figure 4 shows the variation of the maximum value of $\sigma_{\theta\theta}$ with r_2/r_1 for various values of Ω. From these curves it is possible to read off the least volume of reinforcement and the optimum value of r_2/r_1 required to ensure that $\sigma_{\theta\theta}$ does not exceed any specified value.

3. PURE SHEAR AT INFINITY

We now consider Problem II, with the conditions (7) as $r \to \infty$. We seek solutions of the form

$$u(r,\theta,z) = \hat{u}(r,z)\cos 2\theta \,, \qquad v(r,\theta,z) = \hat{v}(r,z)\sin 2\theta \,. \tag{29}$$

The approximations we shall make are similar to those which were made for Problem I, and we shall not detail them explicitly as they arise.

3.1 Stress and deformation in the reinforcement

When they are expressed in terms of $\hat{u}$ and $\hat{v}$, the equilibrium equations (12) become

$$\begin{aligned} c'_{22}\left(\frac{\partial^2\hat{u}}{\partial r^2}+\frac{1}{r}\frac{\partial\hat{u}}{\partial r}\right) - (E_L+4\mu_L)\frac{\hat{u}}{r^2}+\mu_T\frac{\partial^2\hat{u}}{\partial z^2} \\ +2(c'_{12}+\mu_L)\frac{\partial\hat{v}}{\partial r}+2(E_L+\mu_L)\frac{\hat{v}}{r^2} = 0\,, \\ -2(c'_{12}+\mu_L)\frac{1}{r}\frac{\partial\hat{u}}{\partial r}-2(E_L+\mu_L)\frac{\hat{u}}{r^2}+\mu_L\left(\frac{\partial^2\hat{v}}{\partial r^2}+\frac{1}{r}\frac{\partial\hat{v}}{\partial r}\right) \\ -(4E_L+\mu_L)\frac{\hat{v}}{r^2}+\mu_L\frac{\partial^2\hat{v}}{\partial r^2} = 0\,. \end{aligned} \tag{30}$$

Following the procedure used in Problem I, we assume that c'_{22}, c'_{12}, μ_L and μ_T may be neglected in comparison to E_L, and that $\hat{u}$ and $\hat{v}$ do not vary rapidly with r, although they may depend strongly on z. Then (30) are approximated by

$$\begin{aligned} \frac{\partial^2\hat{u}}{\partial z^2}-\alpha^2\left(\frac{\hat{u}}{r^2}+\frac{2\hat{v}}{r^2}\right) = 0\,, \\ \frac{\partial^2\hat{v}}{\partial z^2}-2\beta^2\left(\frac{\hat{u}}{r^2}+\frac{2\hat{v}}{r^2}\right) = 0\,, \end{aligned} \tag{31}$$

where

$$\alpha^2 = E_L/\mu_T\,, \qquad \beta^2 = E_L/\mu_L\,. \tag{32}$$

We denote the as yet unknown values of $\hat{u}$ and $\hat{v}$ at the interface $z = 0$ by $\hat{U}(r)$ and $\hat{V}(r)$ respectively, so that

$$\hat{u}(r,0) = \hat{U}(r)\,, \qquad \hat{v}(r,0) = \hat{V}(r)\,, \tag{33}$$

and then the solution of (31) with zero traction conditions on z = h(r) is

$$\hat{u} = \frac{2\{2\mu_T\hat{U}(r)-\mu_L\hat{V}(r)\}\cosh(mh/r)+\mu_L\{\hat{U}(r)+2\hat{V}(r)\}\cosh\{m(h-z)/r\}}{(4\mu_T+\mu_L)\cosh(mh/r)},$$

$$\hat{v} = \frac{-\{2\mu_T\hat{U}(r)-\mu_L\hat{V}(r)\}\cosh(mh/r)+2\mu_T\{\hat{U}(r)+2\hat{V}(r)\}\cosh\{m(h-z)/r\}}{(4\mu_T+\mu_L)\cosh(mh/r)}, \tag{34}$$

where

$$m^2 = \frac{E_L(4\mu_T+\mu_L)}{\mu_T\mu_L} = \alpha^2 + 4\beta^2 . \tag{35}$$

The most significant stress components are then given, approximately, as

$$\sigma_{\theta\theta} = \frac{E_L(\hat{U}+2\hat{V})\cosh\{m(h-z)/r\}\cos 2\theta}{r\cosh(mh/r)},$$

$$\sigma_{rz} = -\frac{E_L(\hat{U}+2\hat{V})\sinh\{m(h-z)/r\}\cos 2\theta}{mr\cosh(mh/r)} \tag{36}$$

$$\sigma_{\theta z} = -\frac{2E_L(\hat{U}+2\hat{V})\sinh\{m(h-z)/r\}\sin 2\theta}{mr\cosh(mh/r)} .$$

3.2 Stress and deformation in the plate

By the same procedure as in Problem I, we express the equilibrium equations in terms of the mean displacements $\bar{u}$ and $\bar{v}$, and identify these approximately with the displacements at the surface z = 0. This gives

$$2(\lambda+\mu)\left(\frac{d^2\hat{U}}{dr^2}+\frac{1}{r}\frac{d\hat{U}}{dr}\right) - 2(2\lambda+3\mu)\frac{\hat{U}}{r^2} + (3\lambda+2\mu)\frac{1}{r}\frac{d\hat{V}}{dr} - (5\lambda+6\mu)\frac{\hat{V}}{r^2} + \frac{\lambda+2\mu}{\mu t}\hat{\sigma}_{rz}(r,0) = 0 ,$$

$$(\lambda+2\mu)\left(\frac{d^2\hat{V}}{dr^2}+\frac{1}{r}\frac{d\hat{V}}{dr}\right) - (17\lambda+18\mu)\frac{\hat{V}}{r^2} - 2(3\lambda+2\mu)\frac{1}{r}\frac{d\hat{U}}{dr} - 2(5\lambda+6\mu)\frac{\hat{U}}{r^2} + \frac{2(\lambda+2\mu)}{\mu t}\hat{\sigma}_{\theta z}(r,0) = 0 , \tag{37}$$

where

$$\sigma_{rz} = \hat{\sigma}_{rz} \cos 2\theta \qquad \text{and} \qquad \sigma_{\theta z} = \hat{\sigma}_{\theta z} \sin 2\theta \,. \tag{38}$$

In the region $r > r_2$, where $\hat{\sigma}_{rz}(r,0) = 0$ and $\hat{\sigma}_{\theta z}(r,0) = 0$, the solution of (37) subject to (7) is

$$\begin{aligned} \hat{U}(r) &= \frac{Qr}{2\mu} + 4(\lambda+2\mu)\frac{F}{r} + \frac{G}{r^3}\,, \\ \hat{V}(r) &= -\frac{Qr}{2\mu} - (\lambda+2\mu)\frac{F}{r} + \frac{G}{r^3}\,, \end{aligned} \tag{39}$$

where F and G are constants which can be eliminated from (39) to give

$$\begin{aligned} &\frac{\hat{U}}{r} - \frac{\hat{V}}{r} + \frac{d\hat{U}}{dr} - \frac{d\hat{V}}{dr} = \frac{2Q}{\mu}\,, \\ &3(\lambda+2\mu)\frac{\hat{U}}{r} + 12(\lambda+\mu)\frac{\hat{V}}{r} + (\lambda+2\mu)\frac{d\hat{U}}{dr} + 4(\lambda+\mu)\frac{d\hat{V}}{dr} = -\frac{(3\lambda+2\mu)Q}{\mu}\,, \end{aligned} \tag{40}$$

and these relations serve as boundary conditions at $r = r_2$.

In the region $r_1 < r < r_2$, $\hat{\sigma}_{rz}(r,0)$ and $\hat{\sigma}_{\theta z}(r,0)$ are given by (36), and we have the conditions $\sigma_{rr} = 0$, $\sigma_{r\theta} = 0$ at $r = r_1$, which give

$$\left.\begin{aligned} &(\lambda+2\mu)\frac{d\hat{U}}{dr} + \lambda\left(\frac{2\hat{V}}{r} + \frac{\hat{U}}{r}\right) = 0\,, \\ &\frac{d\hat{V}}{dr} - \frac{\hat{V}}{r} - \frac{2\hat{U}}{r} = 0\,, \end{aligned}\right\} \quad \text{at } r = r_1 \,. \tag{41}$$

When the values of $\hat{\sigma}_{rz}(r,0)$ and $\hat{\sigma}_{\theta z}(r,0)$ are inserted, (37) take the form

$$\begin{aligned} &2(\lambda+\mu)\left(\frac{d^2\hat{U}}{dr^2} + \frac{1}{r}\frac{d\hat{U}}{dr} - \phi_1^2\frac{\hat{U}}{r^2}\right) + (3\lambda+2\mu)\left(\frac{1}{r}\frac{d\hat{V}}{dr} - \psi\frac{\hat{V}}{r^2}\right) = 0\,, \\ &(\lambda+2\mu)\left(\frac{d^2\hat{V}}{dr^2} + \frac{1}{r}\frac{d\hat{V}}{dr} - \phi_2^2\frac{\hat{V}}{r^2}\right) - 2(3\lambda+2\mu)\left(\frac{1}{r}\frac{d\hat{U}}{dr} + \psi\frac{\hat{U}}{r^2}\right) = 0\,, \end{aligned} \tag{42}$$

where

$$\begin{aligned} \phi_1^2 &= \frac{2(2\lambda+3\mu)mt + \{(\lambda+2\mu)/\mu\}E_L r \tanh(mh/r)}{2(\lambda+\mu)mt}\,, \\ \phi_2^2 &= \frac{(17\lambda+18\mu)mt + 4\{(\lambda+2\mu)/\mu\}E_L r \tanh(mh/r)}{(\lambda+2\mu)mt}\,, \\ \psi &= \frac{(5\lambda+6\mu)mt + 2\{(\lambda+2\mu)/\mu\}E_L r \tanh(mh/r)}{(3\lambda+2\mu)mt}\,. \end{aligned} \tag{43}$$

Equations (42) are solved subject to (41) at $r = r_1$ and to (40) at $r = r_2$. Clearly numerical solution is necessary, but this proves to be quite straightforward. When $\hat{U}$ and $\hat{V}$ have been determined, the stress in the reinforcement is given by (36) and in the plate by (8).

ACKNOWLEDGEMENT

I am grateful to Miss I. J. Arend for carrying out many of the computations.

REFERENCES

[1] McKENZIE, D.O. and WEBBER, J.P.H., *Aero. Quart.* 26 (1975) 254-274

[2] MANSFIELD, E.H., *Int. J. Mech. Sci.* 18 (1976) 469-479

[3] MANSFIELD, E.H., *Royal Aircraft Establishment Report* 76148 (1976)

VIII

ELASTIC WAVE PROPAGATION IN STRONGLY ANISOTROPIC SOLIDS

D. F. PARKER
Department of Theoretical Mechanics
University of Nottingham
Nottingham, NG7 2RD
England

1. BODY WAVES

1.1 Introduction

In many branches of mechanics the study of waves gives much guidance to the understanding of unsteady phenomena. For example, in linear theories unsteady disturbances are commonly represented as superpositions of plane harmonic waves. A study of such waves reveals many important directional properties of the carrying medium - and the existence of an imaginary propagation speed is frequently taken to imply instability. For strongly anisotropic solids the propagation speed depends greatly on the direction of propagation, so that spreading wavefronts are far from spherical. This chapter will deal with some important features of the linear theory and will relate them to predictions for an ideally constrained solid. Some generalisations to acceleration waves in finite elasticity are also mentioned.

Both the above theories deal with body waves. Previous chapters have shown the widespread occurrence and crucial importance of surface layers of concentrated stress. Such layers are also important for the propagation of waves near boundaries. Section 2 will show how strong anisotropy affects the linear theory of waves in plates, and the related theory of Rayleigh waves. Also, some results for non-linear bending waves in plates will be given.

In linear theory, the Euler equation may be taken as

$$\sigma_{ij,j} = \rho_0 \ddot{u}_i \tag{1}$$

with the Cauchy stress $\sigma_{ij} = C_{ijk\ell} e_{k\ell}$ related to the infinitesimal stress $e_{ij} = \frac{1}{2}(u_{i,j} + u_{j,i})$ by (10) or (36) of Chapter I for materials having one or two families of fibres, respectively. If one or more of the constraints

$$\text{Incompressibility:} \qquad e_{ii} = 0 \tag{2}$$

$$\text{Inextensibility:} \qquad a_i e_{ij} a_j = 0, \quad b_i e_{ij} b_j = 0 \tag{3}$$

is applied, these constitutive laws are replaced by one of the forms (19), (24), (28), (47) or (49) of Chapter I.

In finite elasticity it is convenient to use Lagrangian coordinates and the Piola-Kirchhoff (engineering) stress $\underset{\sim}{\pi} = J\underset{\sim}{\sigma}(\underset{\sim}{F}^T)^{-1}$, where

$$J = \det \underset{\sim}{F} = I_3^{\frac{1}{2}} = \rho_0/\rho \tag{4}$$

is the dilatation. The Euler equations then become

$$\pi_{iA,A} = \frac{\partial \pi_{iA}}{\partial X_A} = \rho_0 \ddot{u}_i \, , \tag{5}$$

where dots denote material time derivatives. The constitutive law (58) of Chapter I simplifies to

$$\pi_{iA} = F_{iR}\left\{\frac{\partial W}{\partial C_{RA}} + \frac{\partial W}{\partial C_{AR}}\right\} = \frac{\partial W}{\partial F_{iA}} \, , \tag{6}$$

with appropriate modifications when one or more of the constraints

Incompressibility: $J = 1$ (7)

Inextensibility: $\underset{\sim}{a}_0.\underset{\sim}{C}.\underset{\sim}{a}_0 = 1, \quad \underset{\sim}{b}_0.\underset{\sim}{C}.\underset{\sim}{b}_0 = 1$ (8)

is imposed.

Equations (1) - (8) are valid for non-uniformly anisotropic solids. However, attention will be confined in this chapter to homogeneous materials (with $\underset{\sim}{a}, \underset{\sim}{b}$ independent of $\underset{\sim}{x}$, or $\underset{\sim}{a}_0, \underset{\sim}{b}_0$ independent of $\underset{\sim}{X}$). Then none of the equations (1) - (3), or (5) - (8), defines a natural scale of length or time (e.g. (1) is homogeneous in second-order derivatives of $\underset{\sim}{u}$, (5) may be rearranged as

$$\pi_{iA,A} = \rho_0 \dot{v}_i, \qquad \dot{F}_{iA} = v_{i,A} \tag{9}$$

which is homogeneous in first order derivatives of $\underset{\sim}{F}$ and the velocity $\underset{\sim}{v} = \dot{\underset{\sim}{u}}$, while none of (2), (3), (7) and (8) involves any derivatives of either $\underset{\sim}{e}$ or $\underset{\sim}{F}$). Neither the linear nor the non-linear constitutive laws considered in Chapter I predict that body waves exhibit *frequency dispersion* - the *wave speeds cannot depend on wavelength*. Consequently these constitutive laws do not apply to disturbances having wavelength comparable to the fibre spacing. Theories (see, for example [1,2]) predicting dispersive and scattering effects will not be discussed in this chapter.

1.2 Linear Theory

A general treatment of linear elastic waves in homogeneous anisotropic solids will be found in Musgrave [3]. Strongly anisotropic solids fit within this theory, while constrained materials are exceptional limiting cases. For this reason we first outline some results for unconstrained materials.

It is common to seek solutions of the Euler equations

$$\sigma_{ij,j} = C_{ijk\ell}u_{k,\ell j} = \rho \ddot{u}_i \qquad \text{(writing } \rho_0 = \rho\text{)} \tag{10}$$

in the form

$$u_k = Ap_k \cos(k_r x_r - \omega t) = Ap_k \cos\omega(s_r x_r - t) = Ap_k \cos\phi \tag{11}$$

where A = amplitude (> 0), $\mathbf{p}$ = unit displacement vector, $\mathbf{k}$ = wave vector, ω = angular frequency, $\mathbf{s} = \omega^{-1}\mathbf{k}$ = slowness vector and $\phi \equiv \mathbf{k}.\mathbf{x} - \omega t$ is the phase variable. Displacements (11) describe periodic wavetrains in which the phase planes ϕ = constant advance in the direction of the unit normal $\mathbf{n} = \mathbf{k}/|\mathbf{k}|$ at speed $v = \omega/|\mathbf{k}| = |\mathbf{s}|^{-1} = -\phi_t/|\nabla\phi|$. They provide solutions to (10) for arbitrary A and ω provided that

$$(C_{ijk\ell}s_j s_\ell - \rho\delta_{ik})p_k = 0 \tag{12}$$

or equivalently

$$(C_{ijk\ell}n_j n_\ell - \rho v^2\delta_{ik})p_k = 0. \tag{13}$$

For arbitrary choice of wavenormal $\mathbf{n}$ the (symmetric) *acoustic tensor* $\boldsymbol{\gamma}$ with components $\gamma_{ik} = C_{ijk\ell}n_j n_\ell$ has three real positive eigenvalues ρv^2 (provided that the material satisfies the strong ellipticity condition $m_i n_j C_{ijk\ell} m_k n_\ell > 0 \;\; \forall\; \mathbf{m},\mathbf{n} \neq \mathbf{0}$). The corresponding (mutually orthogonal) unit eigenvectors $\mathbf{p}$ are the displacement vectors associated with planes advancing in the direction $\mathbf{n}$, with phase speed $\pm v = \pm v(\mathbf{n})$, where

$$S(\mathbf{n},v) \equiv \det(C_{ijk\ell}n_j n_\ell - \rho v^2\delta_{ik}) = 0. \tag{14}$$

In an isotropic material, the largest speed v always has $\mathbf{p}$ parallel to $\mathbf{n}$ (longitudinal waves) whilst the remaining speeds coincide and give $\mathbf{p}$ orthogonal to $\mathbf{n}$ (transverse waves). In general materials, plane waves are neither purely longitudinal (*dilatational*) nor transverse (*shear*) except, for example, when $\mathbf{n}$ is parallel to a principal axis of material symmetry (e.g. $\mathbf{n} \parallel$ to $\mathbf{a}$ in (10) of Chapter I).

The closed centro-symmetric surface consisting of the tips of the vectors $\pm\mathbf{n}v(\mathbf{n})$ is known as the *velocity surface* of the material. Of greater use for the display of the directional features of wave propagation is the *wave surface*. This is the envelope of phase planes having unit normal $\mathbf{n}$ and which have travelled from an origin O a distance $v\mathbf{n}$ in unit time. Typically, the velocity and wave surfaces each consist of three sheets. They frequently are represented [3] by certain of their principal sections. For materials which are transversely isotropic about some direction $\mathbf{a}$, all surfaces have rotational symmetry about $\mathbf{a}$ and so may be represented by two-dimensional diagrams. These materials include hexagonal crystals (see Chapter 8 of [3]) and materials reinforced by a single family of fibres (c.f. (10) of Chapter I).

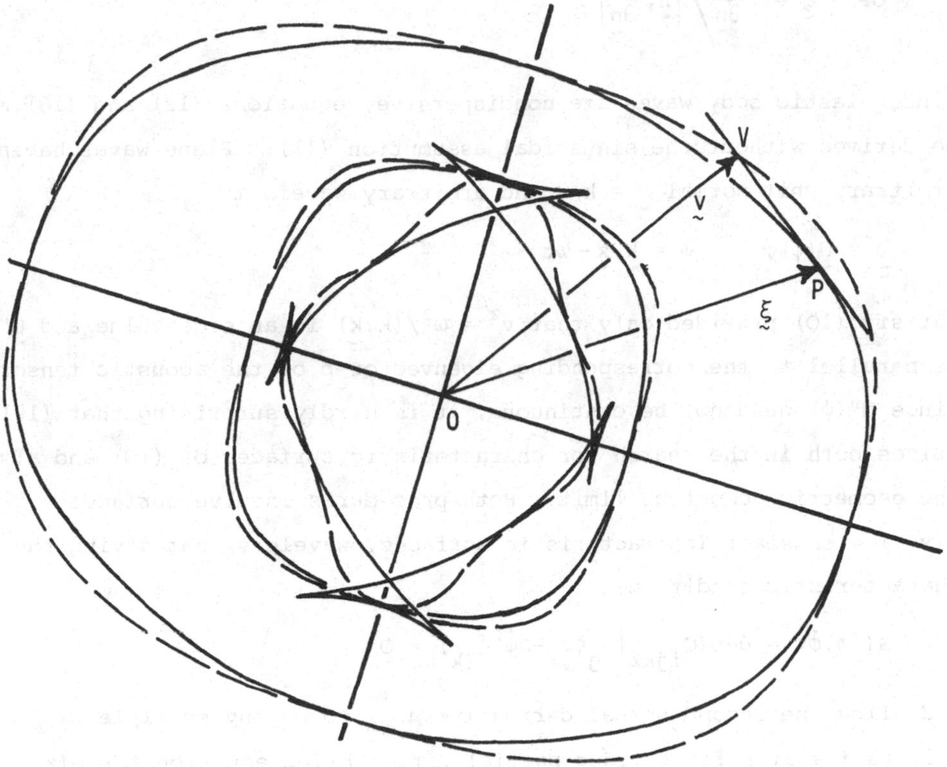

Figure 1. Section through the velocity surface ——— and wave surface ——— of a typical material

Figure 1 shows the relationship between a typical point V (position $\underset{\sim}{n}v(\underset{\sim}{n})$) of the velocity surface and the corresponding point P of the wave surface. A generalisation of Huygens' construction then suggests that signals associated with the displacement vector $\underset{\sim}{p}$ and wave normal $\underset{\sim}{n}$ are transmitted along rays having the direction of $\vec{OP} \equiv \underset{\sim}{\xi}$, the *group velocity* vector. This deduction may be confirmed rigorously:

The envelope in $\underset{\sim}{\xi}$ space of the planes

$$\underset{\sim}{\xi}.\underset{\sim}{n} - v = 0$$

subject to $S(\underset{\sim}{n},v) = 0$ is found, using Lagrange multipliers, to have the form $\underset{\sim}{\xi} = \lambda \partial S/\partial \underset{\sim}{n}$. Use of (15) then defines the ray $\vec{OP}$ as

$$\vec{OP} = \underset{\sim}{\xi} = v \frac{\partial S}{\partial \underset{\sim}{n}} \bigg/ \left(\underset{\sim}{n} . \frac{\partial S}{\partial \underset{\sim}{n}} \right) . \tag{16}$$

Since elastic body waves are nondispersive, equations (12) and (16) may be derived without the sinusoidal assumption (11). Plane waves having arbitrary unit normal $\underset{\sim}{n} = \underset{\sim}{k}/\omega$ and arbitrary waveform

$$\underset{\sim}{u} = \underset{\sim}{U}(\phi), \qquad \phi = \underset{\sim}{k}.\underset{\sim}{x} - \omega t \tag{17}$$

satisfy (10) provided only that $v^2 = \omega^2/(\underset{\sim}{k}.\underset{\sim}{k})$ is an eigenvalue and $\underset{\sim}{U}''(\phi)$ is parallel to the corresponding eigenvector $\underset{\sim}{p}$ of the acoustic tensor $\underset{\sim}{\gamma}$. Since $\underset{\sim}{U}''(\phi)$ need not be continuous, it is hardly surprising that (12) arises both in the search for characteristic surfaces of (10) and also in the geometric acoustics limit. Both procedures involve surfaces $\phi(\underset{\sim}{x},t)$ = constant (characteristic surfaces, wavelets) satisfying the characteristic condition

$$S(\underset{\sim}{\nabla}\phi,\dot{\phi}) = \det(C_{ijk\ell}\phi_{,j}\phi_{,\ell} - \rho\dot{\phi}^2\delta_{ik}) = 0 \tag{18}$$

and allow the second normal derivative $\underset{\sim}{u}_{,\phi\phi}$ to be any multiple of $\underset{\sim}{p}$. Now, (18) is itself a first-order partial differential equation for $\phi(\underset{\sim}{x},t)$. It may be integrated [4] along its characteristics

$$dx_i \bigg/ \frac{\partial S}{\partial \phi_{,i}} = dt \bigg/ \frac{\partial S}{\partial \dot{\phi}} = \frac{d\phi}{0} = \frac{d\phi_{,j}}{0} = \frac{d\dot{\phi}}{0}$$

which are the 'bicharacteristic strips' for (10). Consequently portions of characteristic surface having orientation $\underset{\sim}{n} = \underset{\sim}{\nabla}\phi/|\underset{\sim}{\nabla}\phi|$ and phase speed $v = -\dot{\phi}/|\underset{\sim}{\nabla}\phi|$ propagate along *straight rays* at the *group velocity*

$$\frac{d\underset{\sim}{x}}{dt} = \frac{\partial S}{\partial(\underset{\sim}{\nabla}\phi)} \bigg/ \frac{\partial S}{\partial \dot{\phi}} = -\frac{\partial S}{\partial \underset{\sim}{n}} \bigg/ \frac{\partial S}{\partial v} = v \frac{\partial S}{\partial \underset{\sim}{n}} \bigg/ \left(\underset{\sim}{n} . \frac{\partial S}{\partial \underset{\sim}{n}} \right) = \underset{\sim}{\xi} . \tag{19}$$

(Here we have applied the Euler identity

$$\underset{\sim}{n} . \frac{\partial S}{\partial \underset{\sim}{n}} + v \frac{\partial S}{\partial v} = 6S = 0$$

to the sixth-degree homogeneous function $S(\underset{\sim}{n},v)$.) Furthermore, using standard arguments [3] it can be shown that $\underset{\sim}{\xi}$ is the velocity of mean energy flux at points where the wave normal is $\underset{\sim}{n}$.

To simplify (19), it is found useful to regard (14) as defining the *slowness surface*

$$\hat{S}(\underset{\sim}{s}) \equiv S(\underset{\sim}{s},1) = 0, \tag{20}$$

which is a closed centro-symmetric sixth-degree surface in the space of $\underset{\sim}{s} = \underset{\sim}{n}/v$. Rearranging (15) as $\underset{\sim}{\xi}.\underset{\sim}{s} = 1$, we find that (19) becomes

$$\underset{\sim}{\xi} = \frac{\partial \hat{S}}{\partial \underset{\sim}{s}} \Bigg/ \left(\underset{\sim}{s}.\frac{\partial \hat{S}}{\partial \underset{\sim}{s}}\right). \tag{21}$$

This shows that, at each point Q of the slowness surface (20) with $\vec{OQ} = \underset{\sim}{s}$, the normal has the same direction as the group velocity $\underset{\sim}{\xi}$ of waves with wave normal $\underset{\sim}{n} = \underset{\sim}{s}/|\underset{\sim}{s}|$. Furthermore, there is a duality in this result. If $W(\underset{\sim}{\xi}) = 0$ denotes the wave surface composed of the points $\underset{\sim}{\xi}$ given by (16), then the envelope in $\underset{\sim}{s}$ space of planes $\underset{\sim}{s}.\underset{\sim}{\xi} = 0$ subject to $W(\underset{\sim}{\xi}) = 0$ is found to be

$$\underset{\sim}{s} = \frac{\partial W}{\partial \underset{\sim}{\xi}} \Bigg/ \left(\underset{\sim}{\xi}.\frac{\partial W}{\partial \underset{\sim}{\xi}}\right). \tag{22}$$

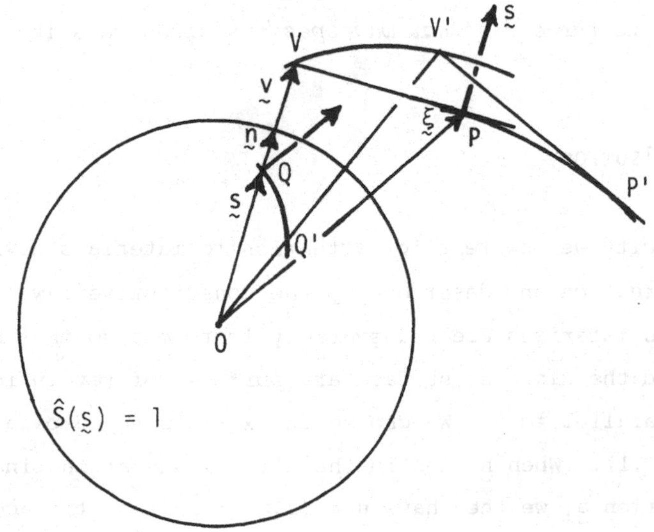

Figure 2. Geometrical relationships between the group velocity $\underset{\sim}{\xi}$, phase velocity $\underset{\sim}{v} = v\underset{\sim}{n}$ and slowness vector $\underset{\sim}{s}$

Equations (21) and (22) are often interpreted (see [3]) as stating that the *slowness surface* $\hat{S}(\underset{\sim}{s}) = 0$ and *wave surface* $W(\underset{\sim}{\xi}) = 0$ are polar reciprocal. Figure 2 shows the geometrical relationship between corresponding points P, V and Q of the wave surface $W(\underset{\sim}{\xi}) = 0$, the velocity surface (14) and the slowness surface $\hat{S}(\underset{\sim}{s}) = 0$, respectively. Since OV.OQ = 1, the plane drawn through V perpendicularly to OV (the polar image of Q in the unit sphere) touches the wave surface $W(\underset{\sim}{\xi}) = 0$ at P, where $\vec{OP} = \underset{\sim}{\xi}$ = group velocity. Similarly the polar image of P is a plane perpendicular to OP which, by (22), touches $\hat{S}(\underset{\sim}{s}) = 0$ at Q. Certain properties of the slowness surface predict important features concerning wave propagation.

The slowness surface has three sheets, with the smaller values of $|\underset{\sim}{s}|$ corresponding to the larger values of $v = |\underset{\sim}{s}|^{-1}$. For most elastic materials it is found [3] that the inner sheet is detached from the other two. Since no straight line can intersect the sixth-degree surface $\hat{S}(\underset{\sim}{s}) = 0$ in more than six points, a detached inner sheet must be convex. Correspondingly there is a single fast wave in the direction of each ray $\underset{\sim}{\xi}$. We shall find that for strongly anisotropic materials one of the other sheets of $\hat{S}(\underset{\sim}{s}) = 0$ is not convex, so that the corresponding sheet of the wave surface has folds along cuspidal edges. This *single* sheet then gives rise to *three distinct wavespeeds* v along rays in many directions $\underset{\sim}{\xi}$.

1.3 Transverse Isotropy

For simplicity we now restrict attention to materials having a single fibre direction and described by the constitutive law (10) of Chapter I. Such materials are transversely isotropic so that both the wave surface and the slowness surface are surfaces of revolution, having symmetry axis parallel to $\underset{\sim}{a}$. We choose the x_3-axis to be parallel to $\underset{\sim}{a}$, so that $\underset{\sim}{a} = (0,0,1)$. When $\underset{\sim}{n}$ lies in the x_1,x_3 plane at inclination θ to the fibre direction $\underset{\sim}{a}$, we then have $\underset{\sim}{n} = (s,0,c)$ so that the acoustic tensor $\underset{\sim}{\gamma} = \underset{\sim}{n}.\underset{\sim}{C}.\underset{\sim}{n}$ is

$$\underset{\sim}{\gamma} = \begin{pmatrix} \mu_L c^2 + (\lambda+2\mu_T)s^2 & 0 & (\lambda+\alpha+\mu_L)cs \\ 0 & \mu_L c^2 + \mu_T s^2 & 0 \\ (\lambda+\alpha+\mu_L)cs & 0 & (\lambda+2\alpha+4\mu_L-2\mu_T+\beta)c^2 + \mu_L s^2 \end{pmatrix} \tag{22}$$

where $s = \sin\theta$, $c = \cos\theta = \underset{\sim}{n}.\underset{\sim}{a}$.

Waves travelling in the fibre direction ($c = 1$, $\theta = 0$) may travel at speed $v = c_1 \equiv \{(\lambda+2\alpha+4\mu_L-2\mu_T+\beta)/\rho\}^{\frac{1}{2}}$ (dilatational waves) or at speed $v = c_2 \equiv (\mu_L/\rho)^{\frac{1}{2}}$ (shear waves). In directions normal to the fibres ($c = 0$, $\theta = \frac{1}{2}\pi$) the wave speeds are $v = c_3 \equiv \{(\lambda+2\mu_T)/\rho\}^{\frac{1}{2}}$ (dilatational waves), $v = c_4 \equiv (\mu_T/\rho)^{\frac{1}{2}}$ (transverse shear waves) and $v = c_2$ (axial shear waves). For general values of θ, one root of the characteristic equation $\det(\underset{\sim}{\gamma}-\rho v^2\underset{\sim}{I}) = 0$ is

$$\rho v^2 = \mu_L c^2 + \mu_T s^2 \tag{23}$$

so that the slowness components $s_1 = s/v$ and $s_3 = c/v$ satisfy $c_4^2 s_1^2 + c_2^2 s_3^2 = 1$. The corresponding displacement vector is

$$\underset{\sim}{p} = (0,1,0) = \underset{\sim}{a} \wedge \underset{\sim}{n}/|\underset{\sim}{a} \wedge \underset{\sim}{n}| . \tag{24}$$

Thus, for all choices of $\underset{\sim}{n}$, there exists one purely transverse wave (an *azimuthal shear wave*) with speed

$$v = (c_2^2 \cos^2\theta + c_4^2 \sin^2\theta)^{\frac{1}{2}} . \tag{25}$$

The corresponding sheet of the slowness surface is the ellipsoid of revolution

$$E_3: \quad c_4^2(s_1^2+s_2^2) + c_2^2 s_3^2 = 1 \tag{26}$$

(prolate if $\mu_T > \mu_L$, oblate if $\mu_L > \mu_T$).

The remaining wave speeds satisfy

$$(c_2^2c^2+c_3^2s^2-v^2)(c_1^2c^2+c_2^2s^2-v^2) - (\lambda+\alpha+\mu_L)^2c^2s^2 = 0.$$

The corresponding sheets of the slowness surface are found by replacing s^2/v^2 and c^2/v^2 by $s_1^2+s_2^2$ and s_3^2 respectively. When the dimensionless parameters

$$A = \frac{c_1^2}{c_2^2} = \frac{\lambda+2\alpha+4\mu_L-2\mu_T+\beta}{\mu_L}, \qquad B = \frac{c_3^2}{c_2^2} = \frac{\lambda+2\mu_T}{\mu_L},$$

$$C = \frac{\lambda+\alpha+\mu_L}{\mu_L}, \qquad D = \frac{c_4^2}{c_2^2} = \frac{\mu_T}{\mu_L}$$

are introduced, this gives

$$\{B(s_1^2+s_2^2)+s_3^2-c_2^{-2}\}\{s_1^2+s_2^2+As_3^2-c_2^{-2}\} - C^2(s_1^2+s_2^2)s_3^2 = 0. \tag{27}$$

To analyse the form of the quartic surface (27) we introduce the non-dimensional quantities

$$X = c_2^2(s_1^2+s_2^2) = \frac{\mu_L(s_1^2+s_2^2)}{\rho} \geqslant 0, \qquad Y = c_2^2 s_3^2 = \frac{\mu_L s_3^2}{\rho} \geqslant 0,$$

so that (27) becomes the quadratic

$$(BX+Y-1)(X+AY-1) - C^2XY = 0. \tag{28}$$

For each material, equation (28) describes a conic which meets the axis $X = 0$ at $Y = 1, A^{-1}$ and the axis $Y = 0$ at $X = 1, B^{-1}$. For isotropic materials ($B = A$, $C = A-1$, $D = 1$) it reduces to the pair of straight lines $X+Y = A^{-1}$, $X+Y = 1$ which correspond to spheres $|\underset{\sim}{s}| = c_3^{-1}$, $|\underset{\sim}{s}| = c_4^{-1}$ describing dilatational and shear waves respectively. Strongly anisotropic materials have $\beta \gg \mu_L$, so giving $A \gg 1$. We consider three cases:

Strongly anisotropic compressible solids. When λ/μ_L, α/μ_L and μ_T/μ_L are comparable with unity, we have $A \gg 1$ with $B,C,D = O(1)$. Equation (28) then shows that either $Y = O(A^{-1})$ or else $BX+Y \simeq 1$. To investigate behaviour near $Y = 0$ we introduce a scaled variable $Z = AY$, so obtaining

$$(BX-1)(X+Z-1) = A^{-1}Z\{(C^2-1)X-Z+1\}. \tag{29}$$

Thus, for $Y = O(A^{-1})$ we have either $X \simeq B^{-1} \simeq B^{-1}(1-Y)$ or $Y \equiv A^{-1}Z = A^{-1}(1-X)$. Consequently, for $A \gg 1$, the slowness sheets lie close to the *intersecting* ellipsoids

$$E_1: \quad B(s_1^2+s_2^2) + s_3^2 = c_2^{-2}, \tag{30}$$

$$E_2: \quad s_1^2 + s_2^2 + As_3^2 = c_2^{-2}, \tag{31}$$

which are defined by the straight lines BX+Y = 1 and X+AY = 1, respectively. By rescaling X and Z we show that the two sheets never intersect.

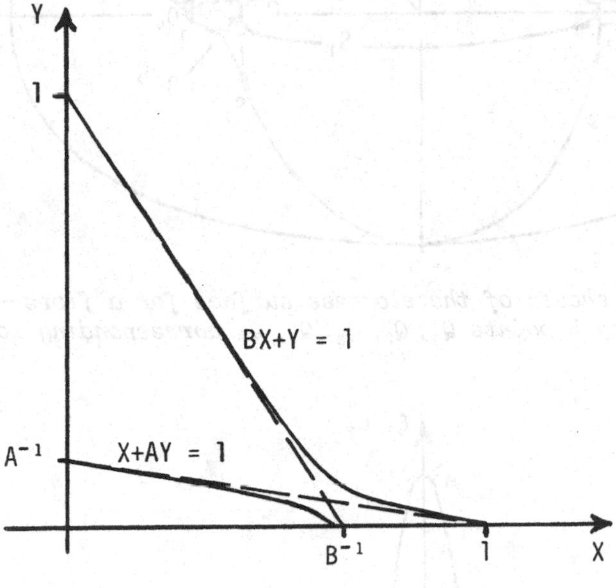

Figure 3

We substitute $X = B^{-1}+A^{-\frac{1}{2}}\bar{X}$, $Z = 1-B^{-1}+A^{-\frac{1}{2}}\bar{Z}$ into (29) and deduce that the hyperbola $\bar{X}(\bar{X}+\bar{Z}) = C^2B^{-3}(B-1)$ gives a good approximation to the transition between the two straight lines. This transition takes place in some interval $X-B^{-1} = O(A^{-\frac{1}{2}})$, as in Figure 3. We deduce that the slowness surface and wave surface have the forms shown in Figures 4 and 5, where A, B, C and D are appropriate to the fibre-epoxy resin composite (see Chapter I). In Figure 4 the outer sheet S_3 is the ellipse E_3 of (26). At all inclinations ψ of the ray $\underset{\sim}{\xi}$ to the fibre direction $\underset{\sim}{a}$, this azimuthal shear wave is the slowest wave (see Figure 5). On the remaining sheets S_2 and S_3, $\underset{\sim}{p}$ lies in the plane of $\underset{\sim}{a}$ and $\underset{\sim}{n}$. It has components $p_{//}$ and $p_{\perp}$ parallel and perpendicular to $\underset{\sim}{a}$ respectively, where (13) and (22) give

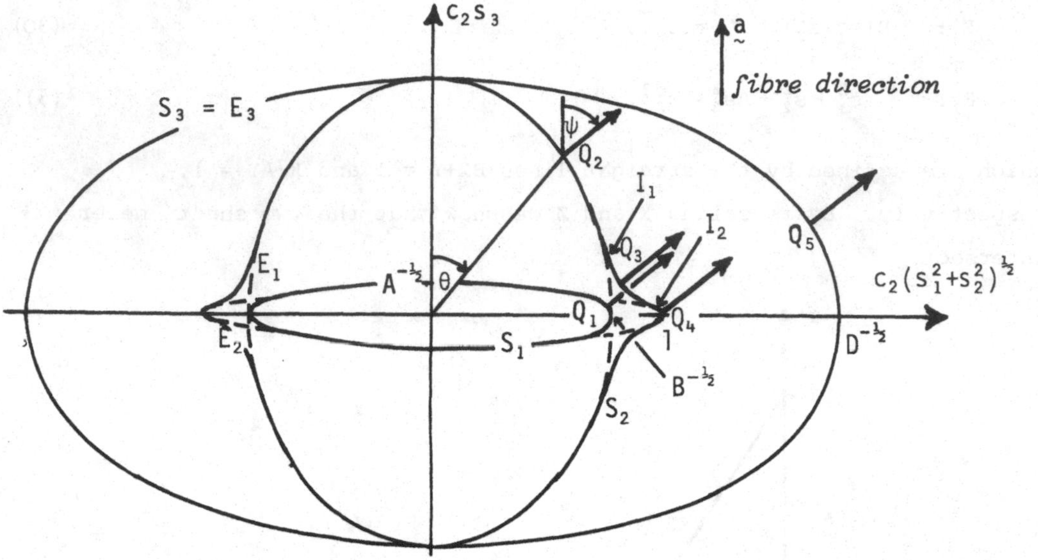

Figure 4. Three sheets of the slowness surface for a fibre-epoxy resin composite, showing 5 points Q_1, Q_2, Q_3, Q_4, Q_5 corresponding to ray vector ξ

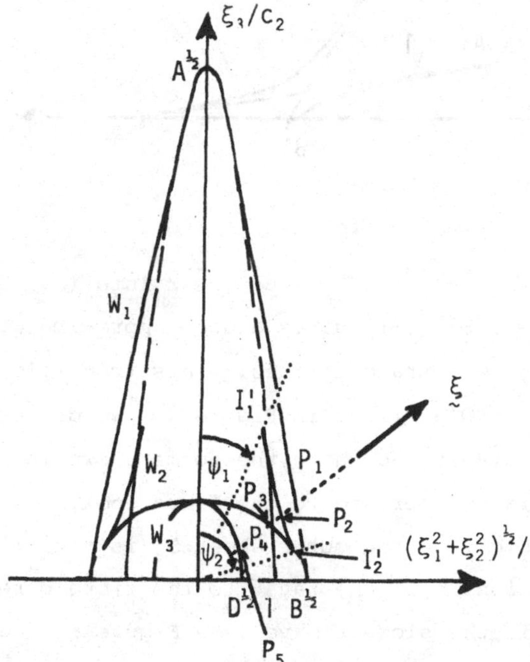

Figure 5. Upper halves of corresponding wave surface sheets showing 5 wave arrivals P_1, P_2, P_3, P_4, P_5

$$\frac{p_{/\!/}}{p_\perp} = -\left(\frac{X}{Y}\right)^{\frac{1}{2}} \frac{BX+Y-1}{CX} = -\left(\frac{X}{Y}\right)^{\frac{1}{2}} \frac{CY}{X+AY-1}, \qquad \left(\frac{X}{Y}\right)^{\frac{1}{2}} = \tan\theta .$$

Consequently, points near the ellipse E_1 have $|p_{/\!/}| \ll |p_\perp|$ while points near ellipse E_2 have $|p_{/\!/}| \gg |p_\perp|$. For most wave inclinations θ, the fastest (quasi-longitudinal) wave propagates with group velocity $\underset{\sim}{\xi} \simeq A^{\frac{1}{2}}\underset{\sim}{a}$ ($\psi \ll 1$) and carries displacements $\underset{\sim}{p} \simeq \underset{\sim}{a}$ (see sheets S_1 and W_1 in Figures 4 and 5). The intermediate sheet S_2 of the slowness surface is re-entrant. If ψ_1 and ψ_2 denote the ray inclinations at the points of inflection I_1 and I_2, then W_2 has three points at each angle ψ satisfying $\psi_1 < \psi < \psi_2$. Consequently, along any ray in the wide cone of angles $\psi_1 < \psi < \psi_2$ an observer should be able to detect *five* wave arrivals, corresponding to P_1 on the wave sheet W_1, points P_2, P_3 and P_4 on the sheet W_2 and, finally, P_5 on sheet W_3. Note also that the axial component $\underset{\sim}{\xi}.\underset{\sim}{a}$ of the group velocity of W_2 waves is considerably greater than $c_2 = (\mu_L/\rho)^{\frac{1}{2}}$ for many points near to the cusp I_1'. We might have anticipated that axial longitudinal waves travel fast, but the fast propagation of a second mode is more surprising.

Strongly anisotropic, almost incompressible solids. When $\lambda/\beta = O(1)$ and $\mu_L, \mu_T \ll \beta$, equations (26) and (28) still hold, with $D = O(1)$ but $A, B, C \gg 1$. The sheets S_1 and S_2 again lie close to E_1 and E_2, although

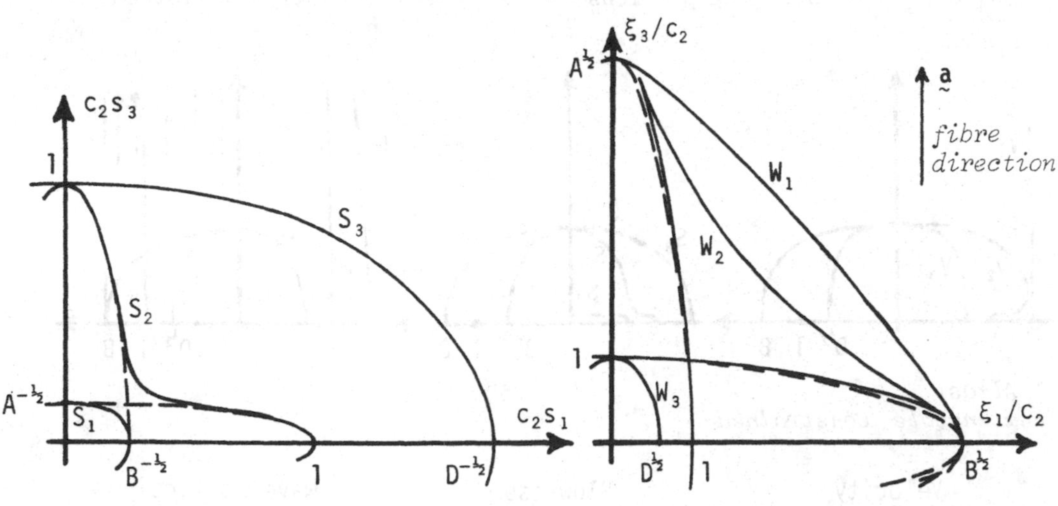

Figure 6

E_1 now is very slender. The slowness and wave surfaces sketched in Figure 6 show that longitudinal waves have large speed for all propagation directions (c.f. slowness and wave surfaces for spruce, a wood which has one low shear modulus, p.123 of [3]).

Constrained solids. Various authors have considered waves in constrained solids [5-8]. The inextensibility constraint $a_i e_{ij} a_j = 0$ gives

$$(\underset{\sim}{a}.\underset{\sim}{p})(\underset{\sim}{a}.\underset{\sim}{n}) = 0 \tag{32}$$

so that a compressible, inextensible material possesses only two finite wavespeeds unless the wave orientation satisfies $\cos\theta = \underset{\sim}{a}.\underset{\sim}{n} = 0$. For $\theta \neq \frac{1}{2}\pi$ the speed corresponding to $\underset{\sim}{p} = \underset{\sim}{a}$ is infinite, not the spurious value $v = 0$ mentioned by Scott [6]. Figure 7 compares the velocity, slowness and wave surfaces for an inextensible material to the limit of the behaviour in Figures 4 and 5 (the arrows indicate that portions of the wave surface lie "at infinity"). Green [7] derives expressions for the three wavespeeds v in powers of $(\mu_L/\beta)^{\frac{1}{2}}$. He draws attention to the singular behaviour for $\cos\theta = O(\mu_L^{\frac{1}{2}}\beta^{-\frac{1}{2}}) = O(A^{-\frac{1}{2}})$, but does not analyse the group velocity $\underset{\sim}{\xi}$. The asymptotically thin groove in sheet V_2 gives rise to the cusps I_1' and I_2'. As $\beta \to \infty$ the point I_1' moves to infinity, giving arbitrarily large group velocity component $\underset{\sim}{\xi}.\underset{\sim}{a}$ for *all* $|\underset{\sim}{\xi} \wedge \underset{\sim}{a}| \leqslant c_3 = B^{-\frac{1}{2}} c_2$. The portions ——·——·—— are easily overlooked

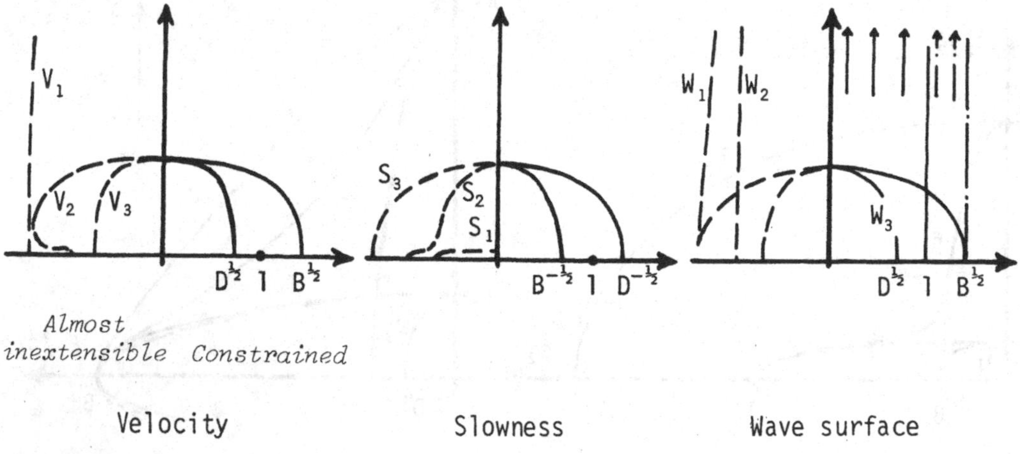

Figur

unless analysis for $\cos\theta = O(A^{-\frac{1}{2}})$ is carefully performed. They arise because *constraints act instantaneously*. For $\underset{\sim}{a} = (0,0,1)$ the inextensibility constraint $u_{3,3} = 0$ is integrable as $u_3 = u_3(x_1,x_2,t)$. The associated shearing motions travel axially *without delay*, and may propagate normal to the fibres at speeds as great as the dilatational speed c_3.

In constrained materials some loadings cause zero displacement, yet the resulting reaction stresses affect the wavespeeds. Chen and Gurtin [5] analyse this and predict that one wavespeed becomes imaginary when a large axial compressible stress is applied, suggesting buckling instability. A similar phenomenon occurs also in Section 2.

For an incompressible (IFRM) material the constraint (2) allows only one finite wavespeed for general orientations θ. Again the constraint may be integrated to give $u_1 = \chi_{,2}$, $u_2 = -\chi_{,1}$ for some $\chi = \chi(\underset{\sim}{x},t)$. In many dynamic problems it is preferable to work with χ and $u_3(x_1,x_2,t)$ directly.

1.4 Acceleration Waves

Nonlinear generalisations are possible, using the Lagrangian formulation. Characteristic surfaces $\phi(\underset{\sim}{X},t)$ = constant of equations (9) satisfy

$$\det(C_{iAkB}\phi_{,A}\phi_{,B}-\rho_0\dot{\phi}^2\delta_{ik}) = 0$$

which is analogous to (18), where $C_{iAkB} = \partial^2 W/\partial F_{iA}\partial F_{kB}$. The 'acoustic tensor' now has components $C_{iAkB}N_A N_B$ which depend on the deformation $\underset{\sim}{F}$ as well as the components of the 'normal' $\underset{\sim}{N}$ to ϕ = constant. Correspondingly the 'slowness' $\underset{\sim}{s}$ depends on the disturbance carried. Waves cannot be superposed but exact results for the propagation of acceleration discontinuities may be derived, even for constrained materials [5,6]. However, these results presuppose knowledge of conditions ahead of the discontinuity. This is unrealistic for inextensible materials except when $\underset{\sim}{N}.\underset{\sim}{a}_0 = 0$. Since, in this case, linear theory suggests that the group velocity is not unique, the analysis seems to have little applicability except when applied to plane waves. Some analytic details may be found in [8].

2. BENDING WAVES IN PLATES

2.1 Linear Theory

This account is based on recent work of my colleague Dr. W. A. Green [9], who considers both inextensible and highly anisotropic materials. For brevity, we consider only waves in a plate $-h \leqslant x_1 \leqslant h$ which is strictly inextensible in the direction $\underset{\sim}{a} = (0,0,1)$, parallel to the x_3-axis. We seek bending waves in which the displacements have the form

$$u_1 = U_1(x_1)\cos\phi\,, \quad u_2 = U_2(x_1)\sin\phi\,, \quad u_3 = U_3(x_1)\sin\phi\,, \tag{33}$$

where the phase variable $\phi \equiv k_2x_2+k_3x_3-\omega t = \underset{\sim}{k}.\underset{\sim}{x}-\omega t$ is constant on planes having unit normal $\underset{\sim}{n} = \underset{\sim}{k}/|\underset{\sim}{k}| = (0, \cos\alpha, \sin\alpha)$ and advancing across the plate with speed $v = \omega/k$. The lateral displacement u_1 is to be an even function of x_1, while u_2 and u_3 must be odd in x_1. The tractions vanish over the lateral surfaces, so that

$$\sigma_{i1} = 0 \quad \text{on} \quad x_1 = \pm h\,. \tag{34}$$

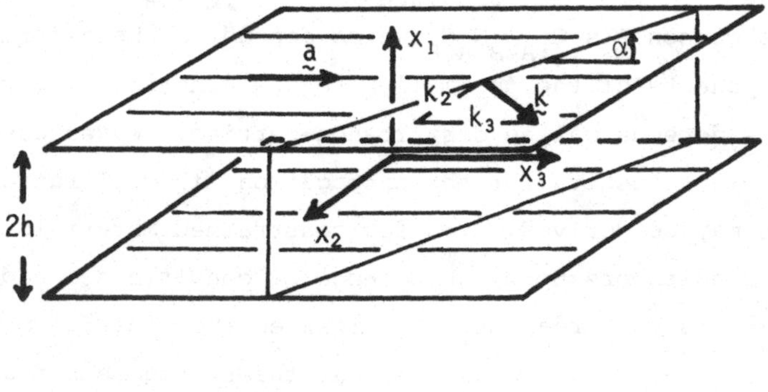

Figure 8. Orientation of wave normal $\underset{\sim}{k}$ *to the fibre direction* $\underset{\sim}{a}$

The inextensibility constraint $u_{3,3} = 0$ gives

$$U_3(x_1) = 0, \quad \text{unless} \quad \sin\alpha = 0. \tag{35}$$

Since the stress may be written, using (25) and (26) of Chapter I, as

$$\sigma_{ij} = (T-\lambda e_{kk})\delta_{i3}\delta_{j3} + \lambda e_{kk}\delta_{ij} + 2\mu_T e_{ij} + 2(\mu_L-\mu_T)(\delta_{i3}e_{3j}+\delta_{j3}e_{3i}),$$

the reaction T to the constraint $e_{33} = 0$ is the total axial stress

$$T = \sigma_{33}. \tag{36}$$

Then, since equations (33) give

$$e_{11} = U_1' \cos\phi, \qquad e_{22} = k_2U_2\cos\phi, \qquad e_{33} = 0,$$

$$e_{23} = \tfrac{1}{2}(k_3U_2+k_2U_3)\cos\phi, \qquad e_{31} = -\tfrac{1}{2}(k_3U_1-U_3')\sin\phi,$$

$$e_{12} = -\tfrac{1}{2}(k_2U_1-U_2')\sin\phi,$$

the stress components are

$$\sigma_{11} = \{(\lambda+2\mu_T)U_1'+\lambda k_2U_2\}\cos\phi, \qquad \sigma_{22} = \{\lambda U_1'+(\lambda+2\mu_T)k_2U_2\}\cos\phi,$$

$$\sigma_{33} = T, \qquad \sigma_{23} = \mu_L(k_3U_2+k_2U_3)\cos\phi,$$

$$\sigma_{31} = -\mu_L(k_3U_1-U_3')\sin\phi, \qquad \sigma_{12} = -\mu_T(k_2U_1-U_2')\sin\phi. \tag{37}$$

The momentum equations (1) reduce to

$$c_3^2U_1'' + (\omega^2-c_4^2k_2^2-c_2^2k_3^2)U_1 + (c_3^2-c_4^2)k_2U_2' = 0,$$

$$-(c_3^2-c_4^2)k_2U_1' + c_4^2U_2'' + (\omega^2-c_3^2k_2^2-c_2^2k_3^2)U_2 = 0,$$

$$T_{,3} = \rho c_2^2k_3(U_1'+k_2U_2)\sin\phi + \rho\{(c_2^2k_2^2-\omega^2)U_3-c_2^2U_3''\}\sin\phi, \tag{38}$$

whilst the boundary conditions (34) become

$$\left.\begin{aligned} c_3^2U_1' + (c_2^2-2c_4^2)k_2U_2 &= 0 \\ k_2U_1 - U_2' &= 0 \\ k_3U_1 - U_3' &= 0 \end{aligned}\right\} \quad \text{at} \quad x_1 = \pm h. \tag{39}$$

The first two of both (38) and (39) form a complete system for $U_1(x_1)$ and $U_2(x_1)$ and will be solved first. The final equations, together with (35), then determine the reaction stress $T(\underset{\sim}{x},t)$ and possible displacement $U_3(x_1)$.

Since we require $U_1(x_1)$ to be even and $U_2(x_1)$ to be odd, we seek solutions of (38) in the form

$$U_1 = P\cosh px_1, \qquad U_2 = Q\sinh px_1$$

for some constants P, Q and p. Substitution then gives

$$(c_3^2p^2+\omega^2-c_4^2k_2^2-c_2^2k_3^2)P + (c_3^2-c_4^2)k_2Q = 0,$$

$$-(c_3^2-c_4^2)k_2P + (c_4^2p^2+\omega^2-c_3^2k_2^2-c_2^2k_3^2)Q = 0,$$

so that p^2 must satisfy the quadratic equation

$$(c_3^2p^2+\omega^2-c_4^2k_2^2-c_2^2k_3^2)(c_4^2p^2+\omega^2-c_3^2k_2^2-c_2^2k_3^2) + (c_3^2-c_4^2)^2k_2^2p^2 = 0.$$

This may be factorised as

$$(c_3^2p^2+\omega^2-c_3^2k_2^2-c_2^2k_3^2)(c_4^2p^2+\omega^2-c_4^2k_2^2-c_2^2k_3^2) = 0,$$

so giving

$$p^2 = k_2^2 + \frac{c_2^2k_3^2-\omega^2}{c_3^2} \equiv p_1^2 \qquad \text{or} \qquad p^2 = k_2^2 + \frac{c_2^2k_3^2-\omega^2}{c_4^2} \equiv p_2^2, \tag{40}$$

with corresponding ratios

$$\frac{P}{Q} = -\frac{p_1}{k_2} \qquad \text{or} \qquad \frac{P}{Q} = -\frac{k_2}{p_2}\,.$$

Consequently, we must have

$$\begin{aligned} U_1(x_1) &= P_1p_1\cosh p_1x_1 - P_2k_2\cosh p_2x_1, \\ U_2(x_1) &= -P_1k_2\sinh p_1x_1 + P_2p_2\sinh p_2x_1. \end{aligned} \tag{41}$$

The corresponding boundary conditions $\sigma_{11} = 0 = \sigma_{21}$ are, from (39),

$$\begin{aligned} &\{c_3^2(p_1^2-k_2^2)+2c_4^2k_2^2\}P_1\sinh p_1h - 2c_4^2k_2p_2P_2\sinh p_2h = 0, \\ &2k_1p_1P_1\cosh p_1h - (k_2^2+p_2^2)P_2\cosh p_2h = 0. \end{aligned} \tag{42}$$

Using the identity

$$c_3^2(p_1^2-k_2^2) = c_4^2(p_2^2-k_2^2) = c_2^2k_3^2 - \omega^2 \tag{43}$$

which follows from (40), the condition that (42) has non-trivial solutions may be written as

$$(p_2^2+k_2^2)^2 \tanh p_1h - 4k_2^2p_1p_2 \tanh p_2h = 0. \tag{44}$$

Dispersion relation. Equation (44) is a secular equation determining the phase speed $v = \omega/(k_2^2+k_3^2)^{\frac{1}{2}} = \omega/k$ in terms of the wavelength $2\pi/k$ and the angle $\alpha = \tan^{-1}(k_3/k_2)$. It may be related to both the secular equation for bending waves in isotropic plates, and to the equation which determines the speed of Rayleigh waves.

For an isotropic material having dilatational and shear wave speeds $\bar{c}_1^2 = (\lambda+2\mu)/\rho$ and $\bar{c}_2^2 = \mu/\rho$, the secular equation is

$$(q_2^2+1)^2 \tanh q_1kh - 4q_1q_2 \tanh q_2kh = 0, \tag{45}$$

where $q_1^2 = 1-v^2/\bar{c}_1^2$, $q_2^2 = 1-v^2/\bar{c}_2^2$. It may be solved in the form

$$v = \bar{c}_2F(kh), \tag{46}$$

where the function $F(kh)$ depends also on the Poisson ratio $\nu = (\tfrac{1}{2}\bar{c}_1^2-\bar{c}_2^2)/(\bar{c}_1^2-\bar{c}_2^2) = \tfrac{1}{2}\lambda/(\lambda+\mu)$. The substitutions $p_1 \to k_2q_1$, $p_2 \to k_2q_2$, $\bar{c}_1 \to c_3$, $\bar{c}_2 \to c_4$, $kh \to k_2h$ convert (45) into (44) and give $v^2 \to (\omega^2-c_2^2k_3^2)/k_2^2$. Consequently, solutions to (44) may be expressed as

$$\omega^2 = c_4^2k_2^2F^2(k_2h) + c_2^2k_3^2. \tag{47}$$

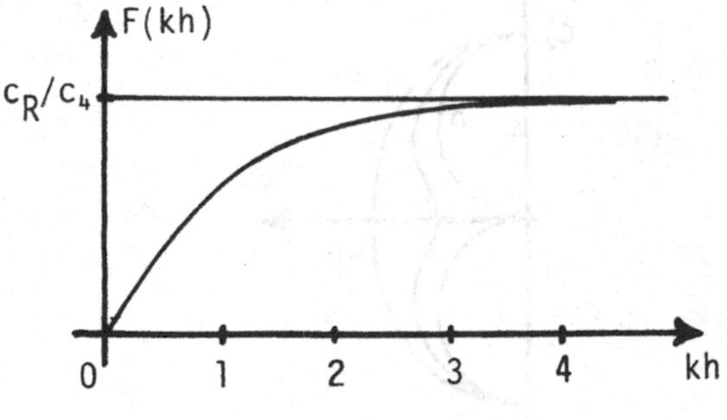

Figure 9. Dimensionless wave speed v/c_4 for waves propagating normal to a

This *dispersion relation* determines possible wave vectors $\underset{\sim}{k} = k(0, \cos\alpha, \sin\alpha)$ for waves having angular frequency ω. (In $F(kh)$ the appropriate value of the Poisson ratio is $\nu = \frac{1}{2}\lambda/(\lambda+\mu_T)$.)

The graph in Figure 9 (based on [10]) shows that the fundamental solution $F(kh)$ to (45) is a monotonically increasing function, with $F(0) = 0$ and $c_4F(kh) \to c_R$ as $kh \to \infty$. Here c_R is the speed of Rayleigh waves in an isotropic solid having moduli λ and μ_T (i.e. the solution $v = c_R$ of the short wave limit $kh \to \infty$ of (45)). From (47) we deduce that the phase speed $v = \omega/k$ of *short waves* ($\lambda \ll \pi h$, i.e. $kh \gg 2$) is given by

$$v^2 \simeq c_R^2 \cos^2\alpha + c_2^2 \sin^2\alpha. \tag{48}$$

Since equation (40) gives $p_1, p_2 = O(k_2)$, displacements in these waves are significant only near the surfaces $x_1 = \pm h$ (the *skin effect*), unless $\cos\alpha$ is small. The limit $kh \to \infty$ corresponds to Rayleigh (surface) waves. Propagation normal to the fibres ($\alpha = 0$) is at the 'isotropic' Rayleigh speed c_R, whilst propagation parallel to the fibres is at the *shear* speed c_2 (as $\alpha \to \frac{1}{2}\pi$, then $p_1, p_2 \to 0$ so that displacements penetrate throughout the plate). By analogy with (25) it can be shown that the 'wave surface' is an ellipse having c_R and c_2 as semi-axes. For *long waves* ($kh \simeq 0$) equation (47) gives $v^2 \simeq c_2^2 \sin^2\alpha$. This predicts vanishing phase speed for waves travelling normal to the fibres (as in isotropic plates) but non-zero speed for other directions. Figure 10 is a sketch of 'velocity surfaces' at typical $h\omega/c_2$.

Since (47) gives $c_2^2 \sin^2\alpha \leqslant v^2 \leqslant c_R^2 \cos^2\alpha + c_2^2 \sin^2\alpha$, waves travelling

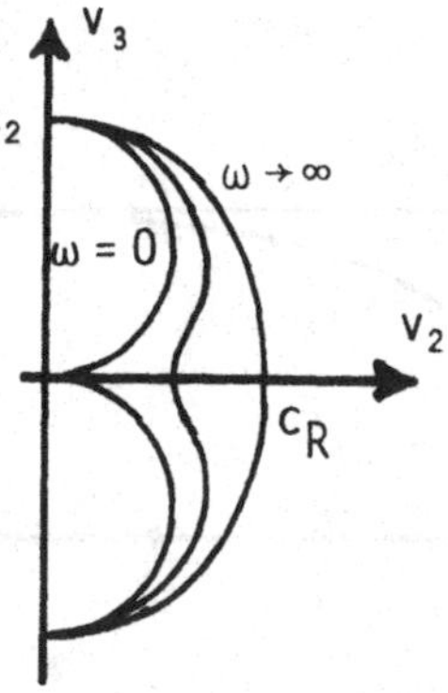

Figure 10. Variation of phase velocity with direction

along the fibres are not dispersed (all wavelengths have phase speed $v = c_2$). In other directions the phase speed varies with wavelength. The associated dispersion may be deduced from (47). The velocity of energy propagation (group velocity, ray velocity) not only differs from the phase velocity

$$\underset{\sim}{v} = \frac{\omega \underset{\sim}{k}}{k^2} = \frac{\{c_4^2 k_2^2 F^2(k_2 h) + c_2^2 k_3^2\}^{\frac{1}{2}}}{k_2^2 + k_3^2}(0, k_2, k_3), \tag{49}$$

but both velocities depend on the angle α and on the dimensionless component $k_2 h$ of the wave vector. Writing $k_2 = \phi_{,2}$, $k_3 = \phi_{,3}$ and $\omega = -\phi_{,t}$, equation (47) becomes

$$G(\phi_{,2}, \phi_{,3}, \phi_{,t}) \equiv c_4^2 \phi_{,2}^2 F^2(h\phi_{,2}) + c_2^2 \phi_{,3}^2 - \phi_{,t}^2 = 0.$$

Like (18) this first-order partial differential equation may be integrated along its 'rays'. This shows that constant values of $\underset{\sim}{k}$ and ω propagate along straight rays, having *group velocity* components

$$\frac{dx_2}{dt} = \frac{c_4^2 k_2 F\{F + hk_2 F'(hk_2)\}}{\omega}, \qquad \frac{dx_3}{dt} = \frac{c_2^2 k_3}{\omega}$$

which depend on the ratio of wavelength to plate thickness through the function $F = F(hk_2)$.

Surface concentration of stress. When $\sin\alpha \neq 0$, (35) gives $U_3 = 0$ and $(38)_3$ gives the reaction stress as

$$T = -\mu_L(U_1' + k_2 U_2)\cos\phi,$$

with $U_1(x_1)$ and $U_2(x_1)$ given by (41) and (42). For $\sin\alpha = 0$, the possibility $U_3 \neq 0$ exists but describes shearing waves $u_1 = u_2 = 0$, not bending waves. These will not be discussed further.

The displacements (41) cannot be chosen to satisfy the boundary conditions $(39)_3$. The explanation is familiar from static analysis (see Chapters II and IV) - *concentrated load* is transmitted by the fibres at the surfaces $x_1 = \pm h$.

We return to the x_3-component of the momentum equation

$$\sigma_{31,1} + \sigma_{32,2} + T_{,3} = \rho\ddot{u}_3. \tag{50}$$

In some layer $h(1-\delta) < x_1 < h$, the stress T becomes singular, but σ_{32} and $\ddot{u}_3$ remain finite. Integrating across this layer gives

$$\Big[\sigma_{32}\Big]_{h(1-\delta)}^{h} + \frac{\partial L}{\partial x_3} = 0, \tag{51}$$

where $L(x_2,x_3,t)$ is the concentrated load. Since (34) gives $\sigma_{31} = 0$ at $x_1 = h$, we obtain from (37)

$$\frac{\partial L}{\partial x_3} = -\mu_L k_3 U_1(h)\sin\phi\,.$$

This shows that the surface layer of fibres must carry the concentrated load

$$L = \mu_L U_1(h)\cos\phi\,. \tag{52}$$

2.2 Finite Amplitude Flexural Waves

Linear theory predicts that bending waves travelling in the fibre direction ($k_3 = 0$, $\cos\alpha = 0$) are non-dispersive, with $v = \mu_L$ for all k. Displacements are independent of x_1 (because $p_1 = p_2 = 0$). These facts allow development of a simple non-linear theory.

Kinematics. Following [11], we decompose the deformation gradient $F_{ij} = \partial x_i/\partial X_j$ in finite plane strain of an incompressible inextensible material (IFRM) as the product

$$\underset{\sim}{F} = \begin{pmatrix} x_{3,3} & x_{3,1} \\ x_{1,3} & x_{1,1} \end{pmatrix} = \begin{pmatrix} \cos\theta & -\sin\theta \\ \sin\theta & \cos\theta \end{pmatrix}\begin{pmatrix} 1 & \gamma \\ 0 & 1 \end{pmatrix} = \underset{\sim}{H}\underset{\sim}{S} \tag{53}$$

of a pure rotation $\underset{\sim}{H}$ and a simple shear $\underset{\sim}{S}$. Similarly we resolve the velocity $\underset{\sim}{v} = \dot{\underset{\sim}{u}}$ into a component u parallel to the fibres and a component v perpendicular to the fibres, writing

$$\underset{\sim}{v} = \underset{\sim}{H}(u,v)^T. \tag{54}$$

As in static deformations, the compatibility conditions $x_{i,jk} = x_{i,kj}$ on the components of $\underset{\sim}{F}$ impose constraints (see Chapter II) which reduce to

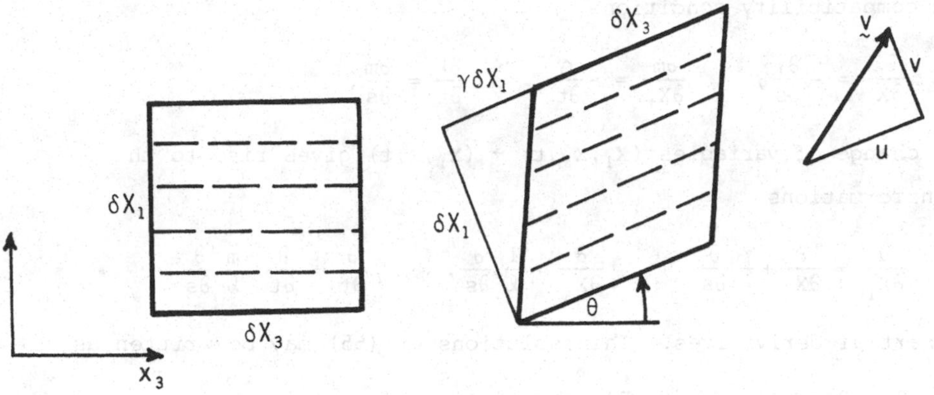

Figure 11. Deformation gradient and velocity referred to the fibre direction

$$\frac{\partial\theta}{\partial X_1} - \gamma\frac{\partial\theta}{\partial X_3} = 0, \qquad \frac{\partial(\gamma-\theta)}{\partial X_3} = 0. \tag{55}$$

Additionally, the compatibility conditions $\dot{F}_{ij} = v_{i,j}$ give

$$\frac{\partial u}{\partial X_1} - v\frac{\partial\theta}{\partial X_1} = \frac{\partial(\gamma-\theta)}{\partial t}, \qquad u\frac{\partial\theta}{\partial X_1} + \frac{\partial v}{\partial X_1} = \gamma\frac{\partial\theta}{\partial t}, \tag{56}$$

$$\frac{\partial u}{\partial X_3} - v\frac{\partial\theta}{\partial X_3} = 0, \qquad u\frac{\partial\theta}{\partial X_3} + \frac{\partial v}{\partial X_3} = \frac{\partial\theta}{\partial t}. \tag{57}$$

Condition $(55)_2$ integrates to give

$$\gamma = \Gamma(X_1,t) + \theta \tag{58}$$

while $(55)_1$ shows that *at each instant* θ is constant along $dX_3/dX_1 = -\gamma$. This means (see Figure 11) that fibre normals are straight (at inclination θ to the x_1-axis). It is convenient to replace X_3 by an intrinsic coordinate s which is constant along fibre normals and measures distance along the central fibre $X_1 = 0$ of the plate $-h \leqslant X_1 \leqslant h$. The coordinate s is *not* a material coordinate. Although we know that the mapping $X_3 = X(X_1,s,t)$ must have

$$\frac{\partial X}{\partial X_1} = -\gamma \qquad \left(= \left.\frac{\partial X_3}{\partial X_1}\right|_{s,t}\right), \tag{59}$$

its remaining partial derivatives $X_{,s} \equiv \ell$ and $X_{,t} \equiv m$ must be found from the compatibility conditions

$$\frac{\partial \ell}{\partial X_1} = -\frac{\partial \gamma}{\partial s}, \qquad \frac{\partial m}{\partial X_1} = -\frac{\partial \gamma}{\partial t}, \qquad \frac{\partial \ell}{\partial t} = \frac{\partial m}{\partial s}. \tag{60}$$

The change of variables $(X_1, X_3, t) \to (X_1, s, t)$ gives rise to the transformations

$$\frac{\partial}{\partial X_1} \to \frac{\partial}{\partial X_1} + \frac{\gamma}{\ell}\frac{\partial}{\partial s}, \qquad \frac{\partial}{\partial X_3} \to \frac{1}{\ell}\frac{\partial}{\partial s}, \qquad \frac{\partial}{\partial t} \to \frac{\partial}{\partial t} - \frac{m}{\ell}\frac{\partial}{\partial s}$$

of partial derivatives. Then solutions of (55) may be written as

$$\theta = \Theta(s,t), \qquad \gamma = \Gamma(X_1,t) + \Theta(s,t). \tag{61}$$

Using these, together with $\ell(0,s,t) = 1$, $m(0,s,t) = 0$ (since $X(0,s,t) = s$), in (60) gives

$$\ell(X_1,s,t) = 1 - X_1\Theta_{,s}, \qquad m(X_1,s,t) = -\psi_{,t} - X_1\Theta_{,t}, \tag{62}$$

where $\psi(X_1,t)$ satisfies $\Gamma = \partial\psi/\partial X_1$, $\psi(0,t) = 0$. Here $\theta = \Theta(x,t)$ is the intrinsic equation of the central fibre $X_1 = 0$, while $\psi(X_1,t)$ is the shearing displacement on any normal $\Theta = 0$.

Equations (57) may be integrated to show that

$$u = U(s,t) + \psi_{,t}, \qquad v = V(s,t), \tag{63}$$

where $U = u(0,s,t)$ and V are the intrinsic velocity components of the central fibre. Equations (56) state that they satisfy

$$\frac{\partial U}{\partial s} - V\frac{\partial \Theta}{\partial s} = 0, \qquad U\frac{\partial \Theta}{\partial s} + \frac{\partial V}{\partial s} = \frac{\partial \Theta}{\partial t}. \tag{64}$$

Equations (61) and (62) state that, at each instant t, the plate occupies one of the *kinematically allowable* configurations $X_3 = s - X_1\Theta(s,t) - \psi(X_1,t)$. Equations (63) and (64) then define the velocity components.

The momentum equations. In plane strain of an IFRM the relevant components of the Cauchy stress $\underset{\sim}{\sigma}$ and Piola Kirchhoff stress $\underset{\sim}{\pi}$ are given by

$$\begin{pmatrix} \sigma_{33} & \sigma_{31} \\ \sigma_{13} & \sigma_{11} \end{pmatrix} = \underset{\sim}{H}\underset{\sim}{\Sigma}\underset{\sim}{H}^T, \qquad \begin{pmatrix} \pi_{33} & \pi_{31} \\ \pi_{13} & \pi_{11} \end{pmatrix} = \underset{\sim}{H}\underset{\sim}{\Sigma}(\underset{\sim}{S}^T)^{-1},$$

where $\underset{\sim}{\Sigma}$ represents the stress components illustrated in Figure 12, namely

$$\underset{\sim}{\Sigma} = \begin{pmatrix} T & G(\gamma) \\ G(\gamma) & -p \end{pmatrix} .$$

Insertion into the Euler equations (5) then yields momentum equations in the fibre and normal directions:

$$\frac{\partial T}{\partial s} - 2G\frac{\partial \Theta}{\partial s} + \ell\frac{\partial G}{\partial X_1} = \rho_0\ell\left(\frac{\partial u}{\partial t} - v\frac{\partial \Theta}{\partial t}\right) , \tag{65}$$

$$\frac{\partial(\ell p)}{\partial X_1} - T\frac{\partial \Theta}{\partial s} - \frac{\partial G}{\partial s} = -\rho_0\ell\left(u\frac{\partial \Theta}{\partial t} + \frac{\partial v}{\partial t}\right) + \rho_0 m\left(u\frac{\partial \Theta}{\partial s} + \frac{\partial v}{\partial s}\right) . \tag{66}$$

Since u, v, ℓ and m are given by (62) and (63), these may be regarded as equations for the variation of the 'tension' T along the fibres and for the variation of 'pressure' p along the normals. This shows that *every kinematically admissible plane-strain motion of an IFRM is also mechanically admissible.*

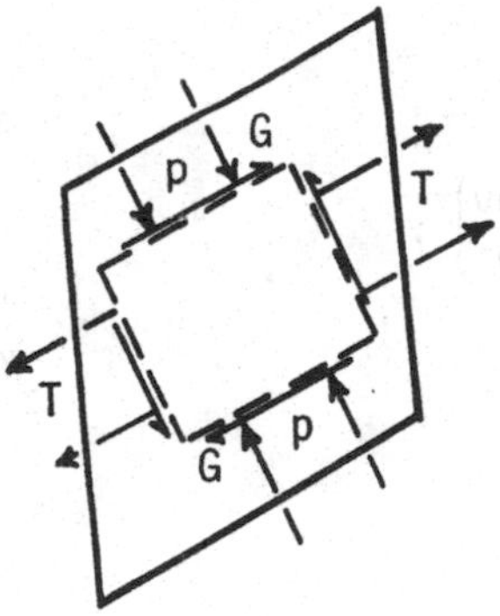

Figure 12. Stresses on a rectangular element

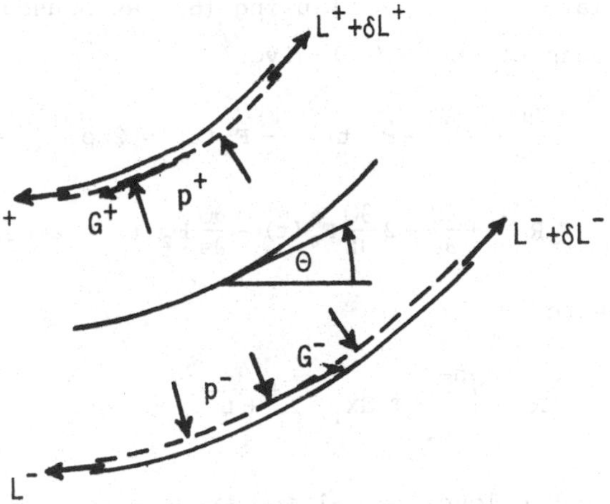

Figure 13. Stresses and loads on the surface layers

Surface layers. As in static deformations, boundary conditions at the surfaces $X_1 = \pm h$ cannot be satisfied unless concentrated loads $L^-(s,t)$ and $L^+(s,t)$ are included at $X_1 = -h$ and $X_1 = h$, respectively. For simplicity, we consider only 'free' portions of plate (no external tractions at $X_1 = \pm h$). If G^- and G^+ are the shear stresses predicted by (61) at $X_1 = -h$ and $X_1 = h$ respectively, we have

$$\frac{\partial L^\pm}{\partial X_3} = \pm G^\pm , \qquad p^\pm = \mp L^\pm \frac{\partial \Theta}{\partial X_3} \qquad \text{at} \quad X_1 = \pm h .$$

These results, derivable intuitively from Figure 13, arise also from boundary layer procedures (see Chapter IV) in the alternative form

$$\left.\begin{aligned} \frac{\partial L^\pm}{\partial s} &= \pm \ell(\pm h, s, t) G^\pm(s,t) \\ \ell(\pm h, s, t) p^\pm &= \mp \Theta_{,s} L^\pm(s,t) , \end{aligned}\right\} \tag{67}$$

where (62) gives $\ell^\pm(h,s,t) = 1 \mp h\Theta_{,s}$. The layers are negligibly thin, so that no inertia contributions arise.

Flexural waves. Equations (65) and (66) may be integrated through the plate ($-h < X_1 < h$) using (67) as boundary conditions. Simplifying by means of (60) - (64) gives

$$\frac{\partial R}{\partial s} - S\frac{\partial \Theta}{\partial s} - \dot{F}_0(t) + \frac{\partial \Theta}{\partial s}\dot{F}_1(t) = 2h\rho_0\left(\frac{\partial U}{\partial t} - V\frac{\partial \Theta}{\partial t}\right) , \tag{68}$$

$$R\frac{\partial \Theta}{\partial s} + \frac{\partial S}{\partial s} - 2\frac{\partial \Theta}{\partial t}F_0(t) - \frac{\partial \Theta}{\partial s}F_2(t) = 2h\rho_0\left(U\frac{\partial \Theta}{\partial t} + \frac{\partial V}{\partial t}\right) , \tag{69}$$

where

$$R \equiv \int_{-h+}^{h-} T \, dX_1 + L^+ + L^- \qquad \text{and} \qquad S = \int_{-h}^{h} G \, dX_1$$

are the longitudinal and transverse components of resultant load at (s,t). The quantities

$$F_0 \equiv \rho_0 \int_{-h}^{h} \frac{\partial \psi}{\partial t} dX_1 , \qquad F_1 \equiv \rho_0 \int_{-h}^{h} X_1 \frac{\partial \psi}{\partial t} dX_1 , \qquad F_2 \equiv \rho_0 \int_{-h}^{h} \left(\frac{\partial \psi}{\partial t}\right)^2 dX_1$$

describing momentum associated with the 'extra shear' $\psi(X_1,t)$ are transmitted *instantaneously* along the fibres.

The constitutive law $G = G(\gamma)$ gives

$$S = \int_{-h}^{h} G(\Theta+\partial\psi/\partial X_1)\, dX_1 \equiv S(\Theta,t), \qquad (70)$$

which reduces to $S = S(\Theta)$ when $\psi \equiv 0$. Inserting (70) into (64), (68) and (69) gives a complete system for U, V, Θ and R. Whenever

$$R + S_{,\Theta} > F_2 - F_0^2/2h\rho_0 \qquad (71)$$

this system possesses the characteristic equations

$$(U+c_2)\,d\Theta + dV = 0 \quad \text{along} \quad ds/dt = c_1\,,$$
$$(U+c_1)\,d\Theta + dV = 0 \quad \text{along} \quad ds/dt = c_2\,, \qquad (72)$$

where the characteristic speeds c_1 and c_2 are

$$c_1,c_2 = [F_0 \pm \{2h\rho_0(R+S_{,\Theta}-F_2)+F_0^2\}^{\frac{1}{2}}]/2h\rho_0\,. \qquad (73)$$

When the inequality (71) is violated, the Cauchy initial value problem is ill-posed, suggesting that instabilities may develop. This agrees with the Kao and Pipkin [12] static buckling criterion $-R > 2hG'(\Theta)$ in the case $\psi \equiv 0$ (giving $S = 2hG(\Theta)$).

In [11], certain exact solutions of (64), (68) and (69) are discussed - for example, simple waves in which U, V, R and Θ are each constant on wavelets given by

$$s - c(\eta)t = \eta\,, \qquad c(\eta) = [\{R+2hG'(\Theta)\}/2h\rho_0]^{\frac{1}{2}} = c_1\,, \qquad (74)$$

with $dR/d\eta = 2hG\,d\Theta/d\eta$, $dU/d\eta = V\,d\Theta/d\eta$, $dV/d\eta = -\{U+c(\eta)\}d\Theta/d\eta$. Since the speed c depends on inclination Θ, the plate profile suffers *amplitude dispersion*. The curvature $\Theta_{,s} = \Theta'(\eta)/\{1+tc'(\eta)\}$ of the central fibre varies with t on each wavelet. If it reaches either value $\Theta_{,s} = \pm h^{-1}$ a *fan* forms, centred on $X_1 = \pm h$, respectively. Subsequently, a fan may travel as a region of uniform angular velocity $\Theta_{,t} = \Omega(t)$ but varying *angle* $s_1(t) < s < s_2(t)$. At the *centre* of the fan, stress concentrations should be expected. Such fans may arise also when a wave is reflected from one edge of the plate. Further details may be found in [11]. Notice that, in *all* deformations some information concerning the disturbance is transmitted instantaneously along the fibres (see $(64)_1$).

REFERENCES

[1] ACHENBACH, J.D.A., *A Theory of Elasticity with Microstructure for Directionally Reinforced Composites*, C.I.S.M. Lectures No. 167, Springer, Vienna, 1975

[2] LEE, E.H., *Dynamics of Composite Materials*, A.S.M.E., New York, 1972

[3] MUSGRAVE, M.J.P., *Crystal Acoustics*, Holden-Day, San Francisco, 1970

[4] GARABEDIAN, P.R., *Partial Differential Equations*, Wiley, New York, 1964

[5] CHEN, P.J. and GURTIN, M.E., *Int. J. Solids Structures* 10 (1974) 275-281

[6] SCOTT, N., *Arch. Rat. Mech. Anal.* 58 (1975) 57-75

[7] GREEN, W.A., *Archives of Mechanics* 30 (1978) 297-307

[8] WHITWORTH, A.M., *Q. Jl. Mech. appl. Math.* 35 (1982) 461-484

[9] GREEN, W.A., *Q. Jl. Mech. appl. Math.* 35 (1982) 485-507

[10] REDWOOD, M.R., *Mechanical Waveguides*, Pergamon, Oxford, 1960

[11] PARKER, D.F., *J. Engng. Maths.* 14 (1980) 57-75

[12] KAO, B.-C. and PIPKIN, A.C., *Acta Mechanica* 13 (1972) 265-280

IX

DYNAMICS OF IDEAL FIBRE-REINFORCED RIGID-PLASTIC BEAMS AND PLATES

A. J. M. SPENCER

Department of Theoretical Mechanics
University of Nottingham
Nottingham, NG7 2RD
England

1. IDEAL FIBRE-REINFORCED BEAMS

1.1 Introduction

The theory described in this chapter is based on references [1] - [9].

We consider a beam reinforced in its longitudinal direction, as shown in Figure 1, and suppose that it undergoes a deformation in which the fibres do not extend. The behaviour of an elastic fibre-reinforced beam under, for example, static point loading, is rather curious. Instead of the deflection illustrated in Figure 2(a), which would occur in an unconstrained isotropic beam, we obtain in the ideal inextensible case the deflection illustrated in Figure 2(b). This phenomenon can easily be demonstrated with a model beam.

The question we consider is how an ideal fibre-reinforced beam would behave in dynamic plasticity. The question is to some extent academic, because real materials are not inextensible. In some situations

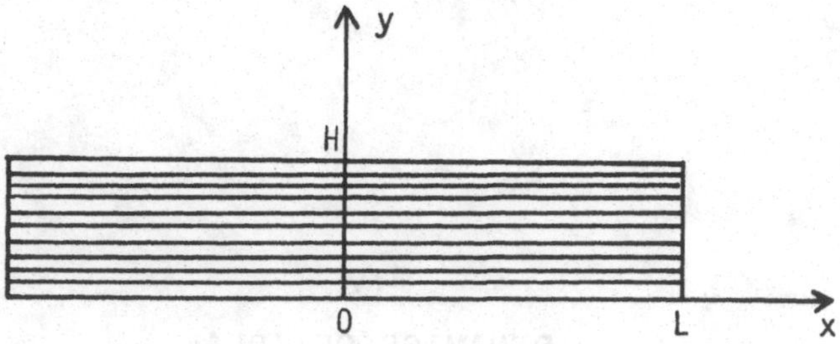

Figure 1. Geometry of ideal fibre-reinforced beam

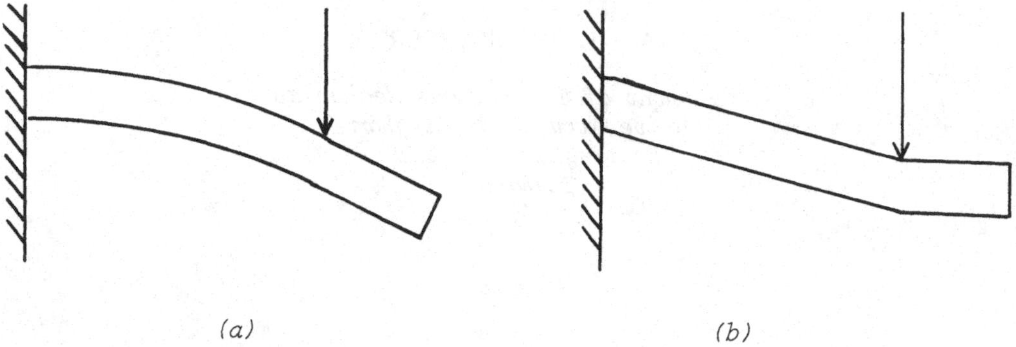

Figure 2. Deflection of (a) an isotropic beam and (b) an ideal fibre-reinforced beam

inextensibility is a good approximation, but in other cases a small amount of extensibility can have a large effect, and this tends to apply to beam and plate problems. For an elastic beam the condition for ideal fibre-reinforced material (IFRM) theory to be acceptable is that

$$\left(\frac{\mu}{E}\right)^{\frac{1}{2}} \frac{2L}{H} \ll 1\,, \tag{1}$$

where μ is a shear modulus, E is the longitudinal extension modulus, L is the length and H is the depth of the beam. This is quite restrictive; even if $\mu/E = 10^{-2}$, which is realistic for some composites, (1) restricts the theory to short, thick beams. For an elastic-plastic beam the equivalent relation to (1) is

$$\left(\frac{\mu_s}{E}\right)^{\frac{1}{2}} \frac{2L}{H} \ll 1\,, \tag{2}$$

where the secant modulus $\mu_s \simeq k/\gamma$, where k is the shear yield stress and γ is the slope of the beam. The condition (2) is more easily satisfied in practice than is (1).

Despite these limitations it is of interest to pursue the theory. It gives interesting qualitative results and insight into the behaviour of fibre-reinforced beams, and is also of mathematical interest because it provides a very simple prototype model for plastic wave propagation.

For non-reinforced rigid plastic beams the main deformation mechanism used is the *plastic hinge*. In its simplest form this theory assumes that the beam remains rigid provided that the bending moment does not exceed a critical bending moment M_0. When $M = M_0$, a plastic hinge forms. Segments on either side of the hinge rotate as rigid bodies. In a dynamic problem the hinge may travel along the beam. The result is that the curvature of the beam is instantaneously discontinuous at the hinge, but the slope is continuous.

Clearly the hinge mechanism is not possible in an IFRM beam, because it involves extension on one side of the beam and contraction on the other, and so violates the inextensibility condition. In fact, in an IFRM beam it is impossible for any segment to rotate rigidly relative to another; a relative rotation must always be accompanied by a shearing deformation in at least one of the segments.

However it is possible for a different type of local discontinuity to propagate. A discontinuity in the slope of the lower edge of the beam leads to a *fan* of the type illustrated in Figure 3. Such a fan, once

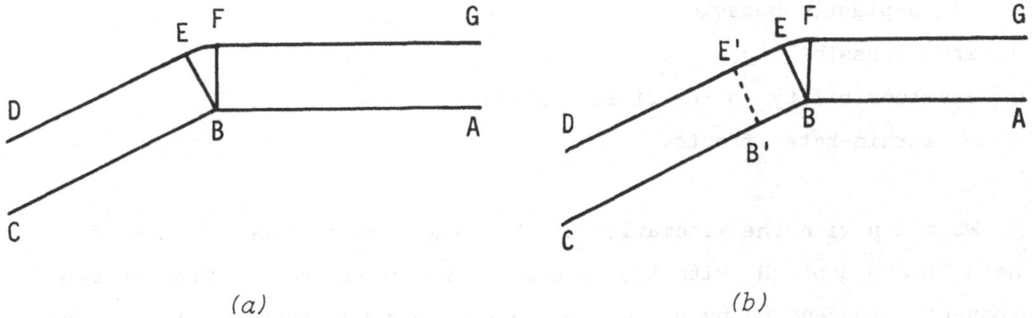

Figure 3. Propagation of a slope discontinuity in an ideal fibre-reinforced beam

formed, may propagate along the beam with material on either side translating as a rigid body. Deformations take place only inside the fan; as the fan passes through a section of the beam, that section undergoes a uniform shear.

It is impossible to form a single fan in the interior of a beam without shear taking place in a region on at least one side of the fan. However, the fan may form at the end of a beam, or two or three adjacent fans may be formed. Some possibilities are illustrated in Figure 4.

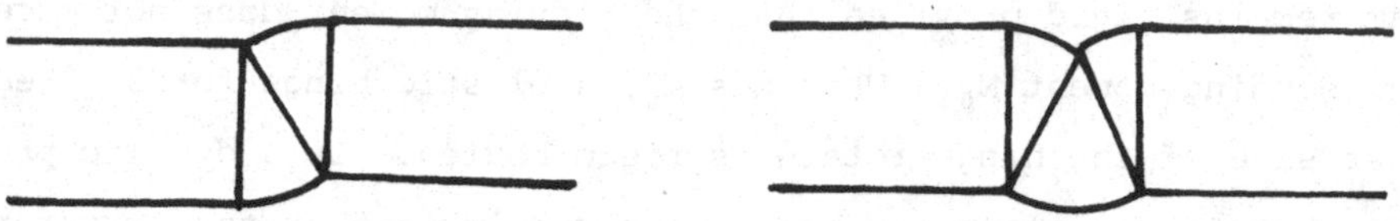

Figure 4. Formation of adjacent centred fan regions in an ideal fibre-reinforced beam

If the slope of the beam is small, the fan angle is small, and the fan can be idealised as a section of the beam across which the slope is discontinuous. This, in fact, is what a small deformation theory permits, and henceforth, for simplicity, we discuss only small deformations.

1.2 General dynamical theory for IFRM beams

We note the main assumptions which will be made. They are:

(a) small deformations,

(b) rigid-plastic behaviour,

(c) incompressibility,

(d) inextensibility in the fibre direction,

(e) no strain-rate effects.

We begin with the kinematics of the beam. We consider a beam of length 2L and depth H, with the axes as shown in Figure 1. Displacement components are denoted by u_x, u_y, u_z, and we assume plane strain, so that $u_z = 0$. Then the incompressibility condition (for small deformations) gives

$$\frac{\partial u_x}{\partial x} + \frac{\partial u_y}{\partial y} = 0\,, \tag{3}$$

and the inextensibility condition gives

$$\frac{\partial u_x}{\partial x} = 0\,. \tag{4}$$

Hence

$$u_x = f(y,t)\,, \qquad u_y = u(x,t)\,. \tag{5}$$

Suppose $u_x = 0$ on some cross-section (for example, a fixed end or an axis of symmetry). Then $u_x = 0$ everywhere, and u_y is the only non-zero displacement component. We denote by $\gamma(x,t)$ the slope of the beam. The only non-zero velocity component is that in the transverse (i.e. y) direction and this is denoted by $v(x,t)$. Hence

$$\gamma(x,t) = \partial u(x,t)/\partial x, \qquad v(x,t) = \partial u(x,t)/\partial t. \tag{6}$$

Clearly $u(x,t)$ must be continuous in x and t, but γ and v may be discontinuous. Suppose that a discontinuity in γ and v propagates along the beam so that at time t it is located at $x = a(t)$. Then continuity of displacement at $x = a(t)$ gives the kinematic condition

$$\dot{a}[\gamma] = -[v]\,, \tag{7}$$

where $\dot{a} = da/dt$ and the square brackets denote the jump in the bracketed variable across the discontinuity.

The equations of motion reduce to

$$\frac{\partial \sigma_{xx}}{\partial x} + \frac{\partial \sigma_{xy}}{\partial y} = 0, \qquad \frac{\partial \sigma_{xy}}{\partial x} + \frac{\partial \sigma_{yy}}{\partial y} = \rho\frac{\partial v}{\partial t}\,. \tag{8}$$

By integrating $(8)_2$ over the cross-sectional area A and remembering that v depends only on x and t, we obtain

$$\frac{\partial Q}{\partial x} + P(x,t) = m\frac{\partial v}{\partial t}\,, \tag{9}$$

where

$$Q(x,t) = \int_A \sigma_{xy} \, dA$$

is the shear force on the cross-section, P(x,t) the resultant force per unit length of beam acting on the lateral surfaces in the y-direction, and $m = \rho A$ is the mass per unit length.

If a discontinuity in v propagates with speed $\dot{a}$, then Q must be discontinuous. The balance of linear momentum gives

$$[Q] = -m\dot{a}[v]. \tag{10}$$

For the deformations under consideration the yield condition reduces to

$$\sigma_{xy}^2 \leqslant k_L^2. \tag{11}$$

Since the shear strain γ is constant on each cross-section of the beam, for homogeneous material the shear stress is likewise constant, so integrating over the cross-section gives

$$Q^2 \leqslant Q_p^2, \tag{12}$$

where $Q_p = kA$ is the magnitude of the critical shear force needed to produce plastic deformation. For positive plastic work Q and γ must have the same sign.

If the material is perfectly plastic, then k is constant and hence Q_p is constant. If the material is strain-hardening, then Q_p depends on the history of γ. For simplicity we consider deformations in which γ is a monotonic function of t at each point of the beam. Then we may write

$$Q_p = Q_p(|\gamma|). \tag{13}$$

For illustration we shall use the specific hardening rule

$$Q_p = Q_0 + Q_1 |\gamma|^n, \tag{14}$$

where Q_0, Q_1 and n are constants. When n = 1 we have linear strain-hardening

$$Q_p = Q_0 + Q_1 |\gamma|. \tag{15}$$

It seems likely that by making suitable choices of Q_0, Q_1 and n, (14) could be made to give a reasonable fit to experimental stress-strain curves.

If the deformation is occurring in a segment of the beam, then $|Q| = Q_p$ in that segment, and the equation of motion (9) becomes

$$\frac{dQ_p}{d|\gamma|}\frac{\partial^2 u}{\partial x^2} - m\frac{\partial^2 u}{\partial t^2} = -P(x,t), \tag{16}$$

so that waves propagate with speed c given by

$$mc^2 = Q_p'(|\gamma|). \tag{17}$$

In the case of linear strain-hardening (15), we denote $c = c_1$ and (17) becomes

$$mc_1^2 = Q_1, \tag{18}$$

and the wave speed c_1 is constant.

More commonly, however, deformation takes place at an isolated section of the beam which instantaneously coincides with a propagating discontinuity. At such a section $Q = \pm Q_p$. For definiteness suppose $Q = +Q_p$. Then (10) becomes

$$[Q_p] = -m\dot{a}[v], \tag{19}$$

and so, from (7) and (10), the speed $\dot{a}$ of the discontinuity is given by

$$m\dot{a}^2 = [Q_p]/[\gamma]. \tag{20}$$

For linear strain-hardening (15) this becomes

$$m\dot{a}^2 = Q_1 = mc_1^2, \tag{21}$$

so that $\dot{a}$ is constant in this case. This feature greatly simplifies the problem in the case of linear strain-hardening.

We note that the equations derived above are formally identical to those which arise in the problem of propagation of plane shear waves in an isotropic rigid-plastic strain-hardening solid.

We have so far made no use of the first equation of motion $(8)_1$. This serves only to determine σ_{xx}, which can be interpreted as the tension in the fibres, and is a reaction to the inextensibility constraint. Since σ_{xy} is independent of y, $(8)_1$ shows that in the interior of the beam σ_{xx} depends only on y and t; its value is determined by end conditions.

The sheets of fibres adjacent to the lateral surfaces may be singular and carry infinite stress but finite force. For a beam of rectangular cross-section this fibre force per unit length in the z-direction is

$$T = \pm \int (Q/A)\, dx + \text{constant}\ , \tag{22}$$

at the upper and lower surfaces.

The present theory uses the critical shear force yield condition (12). The critical bending moment condition, associated with the plastic hinge theory, is

$$M^2 \leqslant M_0^2\ . \tag{23}$$

In general we expect yield to depend on both bending moment and shear force, with a yield condition of the form

$$F(M,Q) \leqslant 0. \tag{24}$$

Thus (12) and (23) can be regarded as two opposite extreme cases of the general condition (24).

1.3 Moving beam brought to rest by central transverse impact

To illustrate the above theory we suppose that a beam moving with speed V_0 in the transverse direction is brought to rest by striking a rigid stop at its mid-point. Details of the solution are given in [1] and [2]. After the impact we assume a configuration of the type illustrated in Figure 5. Slope and velocity discontinuities at $x = \pm a(t)$ propagate outwards from the centre. The central segment $|x| < a(t)$ is at rest; the outer segments $L \geqslant |x| > a(t)$ move as rigid bodies with speed $V(t)$. The speed v and slope γ are discontinuous at $x = a(t)$.

The kinematic condition (7) at $x = a(t)$ is

$$\dot{a}\gamma(a) = V, \tag{25}$$

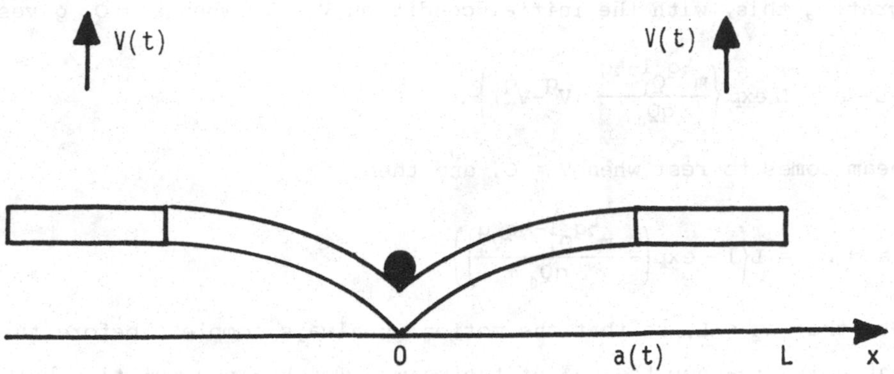

Figure 5. Assumed form of deformation after impact of a beam on a rigid stop

where $\gamma(a)$ denotes the slope at $x = a-0$. The momentum condition (19), with the hardening rule (14), gives

$$m\dot{a}V = Q_1\{\gamma(a)\}^n , \tag{26}$$

and the equation of motion of AB is

$$m(L-a)\dot{V} = -Q_0 . \tag{27}$$

Subtracting (27) from (26) gives

$$-m\frac{d}{dt}\{(L-a)V\} = Q_0 + Q_1\{\gamma(a)\}^n , \tag{28}$$

which is the equation of motion of the entire segment OB.

From (25) and (26)

$$\dot{a} = (Q_1/m)^{1-\frac{1}{2}q}V^{q-1} , \tag{29}$$

where the constant q is introduced to simplify the algebra, and is related to n by

$$q = 2n/(n+1), \qquad n = q/(2-q) . \tag{30}$$

Hence from (27) and (29)

$$\frac{da}{dV} = -\frac{m^{\frac{1}{2}q}Q_1^{1-\frac{1}{2}q}V^{q-1}(L-a)}{Q_0} . \tag{31}$$

Integrating this, with the initial condition $V = V_0$ when $a = 0$, gives

$$L - a = L \exp\left\{\frac{m^{\frac{1}{2}q} Q_1^{1-\frac{1}{2}q}}{qQ_0} (V^q - V_0^q)\right\}. \tag{32}$$

The beam comes to rest when $V = 0$, and then

$$a = a_f = L\left\{1 - \exp\left(-\frac{m^{\frac{1}{2}q} Q_1^{1-\frac{1}{2}q} V_0^q}{qQ_0}\right)\right\}. \tag{33}$$

We note that $a_f < L$, so that the motion is always complete before the discontinuity reaches the end of the beam. To obtain a relation between V and t we substitute for a from (32) into (27) and integrate; to obtain a relation between a and t we substitute for V from (32) into (29) and integrate; and from (25), (26) and (32) we obtain

$$\gamma(a) = \left(\frac{mV^2}{Q_1}\right)^{1-\frac{1}{2}q} = \left\{\left(\frac{mV_0^2}{Q_1}\right)^{\frac{1}{2}q} + \frac{qQ_0}{Q_1} \ln\left(\frac{L-a}{L}\right)\right\}^{(2-q)/q}. \tag{34}$$

The deflection $u(x)$ is given by

$$u(x) = \int_0^x \gamma(x)\, dx .$$

The results are much simpler in the case of linear strain-hardening, $n = 1$, $q = 1$. In this case the solution becomes

$$c_1^2 = Q_1/m , \qquad a = c_1 t ,$$

$$V = V_0 + \frac{Q_0}{c_1 m} \ln\left(1 - \frac{ct}{L}\right), \qquad \gamma(a) = \frac{V_0}{c_1} + \frac{Q_0}{mc_1^2} \ln\left(\frac{L-a}{L}\right),$$

$$u = \frac{V_0 x}{c_1} - \frac{Q_0}{mc_1^2}\left\{x + (L-x)\ln\left(1 - \frac{x}{L}\right)\right\}, \qquad (x \leqslant a). \tag{35}$$

There is further simplification in the case of "small strain-hardening" which arises when $mc_1V_0 \ll Q_0$. In this case

$$a_f \simeq Lmc_1V_0/Q_0 ,$$

and the maximum deflection is

$$u_f \simeq \tfrac{1}{2}LmV_0^2/Q_0 .$$

In this approximation it is easily verified that the initial kinetic energy of the beam, which is mLV_0^2, is equal to the plastic work done in the deformation.

In the case of ideal plasticity, $Q_1 = 0$, and, from (29), $\dot{a} = 0$ and the discontinuity does not propagate. Theoretically the deformation is confined to the point of impact, but of course the solution is no longer valid because the small deformation assumption is violated. A high rate of strain-hardening has the effect of spreading the energy dissipation more evenly along the length of the beam.

1.4 Other impact problems for beams

Various other problems of impact of beams can be treated in a similar way. Some examples are the following:

(a) Impact of a free beam by a mass M moving with speed V_0. This problem is analysed in [3, 8]. The problem described above of a beam striking a rigid stop is equivalent to this problem with $M \to \infty$. Discontinuities propagate from the point of impact but can never reach the free ends of the beam. The problem has an analytical solution but this solution is complicated. Simpler approximate solutions are available in the cases of *(i)* a heavy striker ($M \gg mL$), *(ii)* a light striker ($M \ll mL$), *(iii)* small strain-hardening. The exact solution simplifies considerably in the case of linear strain-hardening.

(b) Impact of a beam supported at its ends by a mass M moving with speed V_0. This problem is analysed in [4, 8]. Again discontinuities propagate outwards from the point of impact, but now if M and V_0 are sufficiently large they may reach the fixed ends and be reflected. An explicit but complicated solution is available for the case in which the discontinuities do not reach the ends; this solution also applies up to the first reflection when reflections do occur. The solution is relatively simple in the case of linear strain-hardening, and has a relatively simple approximation in the case of small strain-hardening. After reflections have taken place, the problem is intractable

analytically except in the case of linear strain-hardening. An interesting result is that for central impact and linear strain-hardening, the slope of the beam behind a reflected discontinuity is uniform.

(c) Impact of a cantilever beam by a mass M moving with speed V_0. This is analysed in [5] for linear strain-hardening; the non-linear case is intractable. Discontinuities propagate from the point of impact; they never reach the free end but for sufficiently large M and V_0 reflection may occur at the fixed end. The case of a cantilever beam struck at its tip is much simpler; it is equivalent to example *(b)* above with impact at the mid-point of the beam.

(d) A clamped beam loaded impulsively so that at time t = 0 the entire span of the beam starts to move with speed V_0. This problem was solved by Laudiero and Jones [9]. In this case discontinuities propagate from the ends of the beam towards the centre; the beam always comes to rest before the discontinuities meet at the centre. A fairly simple analytical solution is obtained for the non-linear hardening rule (14), but as usual the analysis and solution are greatly simplified in the case of linear strain-hardening.

In all of the beam problems discussed in this and the preceding section (and in the plate problems which follow), it can be verified that the rate of plastic working is positive and that the yield condition is not violated in the rigid regions.

A general feature of the solutions is that large values of the strain-hardening constant Q_1 and the strain-hardening index n both have the effect of distributing the deformation, and hence the energy dissipation, more evenly along the length of the beam. For some of the problems it is possible to compare the solutions with the solutions of the corresponding problem for an isotropic beam. For given dimensions and a given shear yield stress, the permanent displacements and response times of the fibre-reinforced beam are significantly smaller (by a factor of order 2H/L) than the displacements and response times of an isotropic beam. Hence fibre-reinforced materials seem to have a considerable potential for weight-saving in energy-absorbing structures.

Laudiero [12] has developed a numerical method of solution of the dynamical theory for fibre-reinforced beams. In test examples, this method gave good agreement with some of the analytical solutions mentioned above.

2. IDEAL FIBRE-REINFORCED PLATES

2.1 General theory

The theory described in this section is based on [6]. We now consider a plate of IFRM material. We have in mind plates of laminated construction, with many laminae, and with each lamina comprising a unidirectionally reinforced thin sheet. Another possibility is in-plane reinforcement by short fibres randomly distributed in planes parallel to the mid-surface of the plate. The mathematical idealisation is that the plate is inextensible in two or more (possibly in all) directions in its plane.

The (x,y) plane is taken as the plane of the mid-surface, w denotes the transverse deflection, and v denotes the transverse velocity component. Suppose one of the directions of inextensibility is inclined at the angle ϕ to the x-axis, so that $\underset{\sim}{a} = (\cos\phi, \sin\phi, 0)$ defines this direction. Then the condition for inextensibility in the direction of $\underset{\sim}{a}$ is

$$\frac{\partial u_x}{\partial x}\cos^2\phi + \left(\frac{\partial u_x}{\partial y} + \frac{\partial u_y}{\partial x}\right)\sin\phi\cos\phi + \frac{\partial u_y}{\partial y}\sin^2\phi = 0\,, \tag{36}$$

where u_x, u_y denote infinitesimal in-plane displacement components.

A solution compatible with any number of constraints of the form (36) and the incompressibility condition is

$$u_x = 0, \qquad u_y = 0, \qquad w = w(x,y,t)\,. \tag{37}$$

For reinforcement in three or more directions (i.e. with constraints of the form (36) with three or more distinct values of ϕ) equation (37) represents the only possible bending solution; for reinforcement in two directions only some other solutions are possible but are not relevant to this theory. The assumption of incompressibility is not essential - most of the analysis is valid for compressible materials if w is interpreted as the deflection of the middle surface and certain stress resultants of relatively small magnitude are neglected.

Since w is independent of z, for homogeneous material the shear stress components σ_{xz}, σ_{yz} are independent of z, and the shear stress resultants are given by

$$Q_x = 2h\sigma_{xz}, \qquad Q_y = 2h\sigma_{yz},$$

where 2h is the plate thickness. We make the mild assumption that the material has sufficient material symmetry for the conditions $e_{xx} = e_{yy} = e_{zz} = e_{xy} = 0$ to imply that the extra-stress stress components s_{xx}, s_{yy}, s_{zz} and s_{xy} (see Chapter I, Section 4) are zero. Then the yield function reduces to a function of $s_{xz} = \sigma_{xz}$ and $s_{yz} = \sigma_{yz}$ or, equivalently, of Q_x and Q_y. Thus the yield condition takes the form

$$F(Q_x,Q_y) = Q_p. \tag{38}$$

We make the major assumption, which was discussed in Section 4 of Chapter I, that Q_p depends on the single parameter W_p, the plastic work, and Q_p is a positive increasing function of W_p. We also assume, without loss of generality, that Q_p has the dimensions of a shear stress resultant. It is further assumed that $F(Q_x,Q_y)$ is a non-negative convex function of its arguments and that $\{F(Q_x,Q_y)\}^2$ is a homogeneous function of degree two. These assumptions are important in the development of the theory.

Any further symmetry will restrict the allowable forms of $F(Q_x,Q_y)$. We do not discuss these restrictions but we mention some special forms of the yield condition which are useful for illustrations of the theory.

(a) $$(2Q_x^2 \cos^2\beta + 2Q_y^2 \sin^2\beta)^{\frac{1}{2}} = Q_p, \qquad (\beta = \text{constant}). \tag{39}$$

This is the case in which F^2 is quadratic in Q_x and Q_y so that (39) is somewhat analogous to von Mises' condition in isotropic plasticity. The yield curve is an ellipse in the Q_x,Q_y plane.

$$\text{(b)} \quad (Q_x^2+Q_y^2)^{\frac{1}{2}} = Q_p . \tag{40}$$

This is the form to which (39) reduces when the material is transversely isotropic about the z-direction. The yield curve becomes a circle.

$$\text{(c)} \quad \begin{aligned} \alpha|Q_x| &= Q_p, \qquad \text{for} \quad \alpha^{-1}|Q_y| < Q_p, \\ \alpha^{-1}|Q_y| &= Q_p, \qquad \text{for} \quad \alpha|Q_x| < Q_p, \end{aligned} \tag{41}$$

where α is constant. The yield curve is piecewise linear (in fact, a rectangle) in the Q_x,Q_y plane. This condition is somewhat analogous to Tresca's condition in isotropic plasticity. If the plate is reinforced in the x and y directions and these directions are mechanically equivalent, then $\alpha = 1$.

For the flow rule we assume as before that the yield function is a plastic potential. We denote

$$\gamma_x = \frac{\partial w}{\partial x}, \qquad \gamma_y = \frac{\partial w}{\partial y}. \tag{42}$$

Then the flow rule is

$$\dot{\gamma}_x = \dot{\bar{\gamma}} \frac{\partial F(Q_x,Q_y)}{\partial Q_x}, \qquad \dot{\gamma}_y = \dot{\bar{\gamma}} \frac{\partial F(Q_x,Q_y)}{\partial Q_y}, \tag{43}$$

where $\dot{\bar{\gamma}}$ is a scalar multiplier.

We then have (using the homogeneity property of $F(Q_x,Q_y)$)

$$\dot{W}_p = Q_x\dot{\gamma}_x + Q_y\dot{\gamma}_y = \dot{\bar{\gamma}}\left(Q_x \frac{\partial F}{\partial Q_x} + Q_y \frac{\partial F}{\partial Q_y}\right) = \dot{\bar{\gamma}} F(Q_x,Q_y) = \dot{\bar{\gamma}} Q_p(W_p) . \tag{44}$$

Hence there is a one-to-one correspondence between W_p and $\bar{\gamma}$; $\bar{\gamma}$ may be interpreted as an equivalent plastic strain and we may regard Q_p as a function of $\bar{\gamma}$. For illustration, and by analogy with (14), we shall take

$$Q_p = Q_0 + Q_1 \bar{\gamma}^n , \tag{45}$$

where Q_0, Q_1 and n are constants. For the yield condition (39) we then find that

$$\dot{\bar{\gamma}} = \left(\frac{\dot{\gamma}_x^2}{2\cos^2\beta} + \frac{\dot{\gamma}_y^2}{2\sin^2\beta} \right)^{\frac{1}{2}} ,$$

and for the yield condition (41) that

$$\dot{\bar{\gamma}} = \alpha^{-1} |\dot{\gamma}_x| + \alpha |\dot{\gamma}_y| .$$

The transverse speed $v = \partial w/\partial t$ and the slopes γ_x and γ_y may be discontinuous across propagating curves in the (x,y) plane. Suppose C is such a curve with outward unit normal $\underset{\sim}{n} = (n_x, n_y, 0)$, and c its speed of propagation in the direction of $\underset{\sim}{n}$. We denote

$$\gamma_n = n_x \gamma_x + n_y \gamma_y = \frac{\partial w}{\partial n} , \qquad Q_n = n_x Q_x + n_y Q_y . \tag{46}$$

Then the jump conditions across C are

$$[v] = -c[\gamma_n] , \qquad [Q_n] = -mc[v] , \tag{47}$$

where m is the mass of unit area of the plate. It follows that

$$mc^2 = [Q_n] / [\gamma_n] . \tag{48}$$

2.2 Impulsive loading of a circular plate

As a first example we investigate the problem of a circular plate, supported at its edge $r = r_0$, and subjected at time t = 0 to a uniformly distributed impulse I per unit area, so that it starts to move with uniform transverse speed $V_0 = I/m$. This problem and a similar problem for a rectangular plate are solved in detail in [7]. We assume the material to be transversely isotropic about the z direction, so that the yield condition (40) is appropriate. We seek solutions in which v and $\gamma_r = \partial w/\partial r$ are discontinuous across a moving circle r = a(t) which propagates *inwards* from $r = r_0$. Each element comes to rest as it is traversed by the discontinuity, so that $a(t) < r \leqslant r_0$ is at rest. The region $0 \leq r < a(t)$ is assumed to move as a rigid body with speed $v(t)$. As the curves C are circles r = constant, we here denote Q_n by Q_r and γ_n by $\gamma_r = \partial w/\partial r$. Then the governing equations are:

(a) Equation of motion of the interior region $r < a(t)$

$$m\pi a^2 \dot{v} = -2\pi a Q_0 \ ;$$

(b) Kinematic jump condition

$$v = \dot{a}\gamma_r \ ;$$

(c) Dynamic jump condition

$$m\dot{a}v = -Q_1\{-\gamma_r(a)\}^n \ .$$

These are three equations for a, v, γ_r, with initial conditions

$$a = r_0 \ , \qquad v = V_0 = I/m \ , \qquad \text{at} \quad t = 0 \ .$$

The solution is, after simplification,

$$\left(\frac{v}{V_0}\right)^q = \frac{\ln(a/a_f)}{\ln(r/a_f)} \ ,$$

$$\gamma_r(a) = -\frac{1}{2q}\left(\frac{V_0}{c_0}\right)^2 \frac{\{\ln(a/a_f)\}^{(2-q)/q}}{\{\ln(r_0/a_f)\}^{2/q}} \ , \tag{49}$$

$$\int_a^{r_0} \left\{\ln\left(\frac{\rho}{a_f}\right)\right\}^{(1-q)/q} d\rho = 2q\left\{\ln\left(\frac{r_0}{a_f}\right)\right\}^{1/q} c_0^2 t/V_0 \ ,$$

where

$$q = 2n/(n+1) \ , \qquad c_0^2 = Q_0/m \ , \qquad c_1^2 = Q_1/m \ ,$$

and

$$a_f = r_0 \exp\left(-\frac{1}{2q}\frac{c_1^{2-q}}{c_0^2} V_0^q\right) \ ,$$

is the value of a at which the plate comes to rest. Note that $a_f > 0$, so the discontinuity never reaches the centre. The solution simplifies in the case of q = 1 of linear strain-hardening.

Graphs of the final deflection for various values of the parameters

q, V_0/c_0, a_f/r_0 are given in [7]. It is shown there that the maximum final deflection $\hat{w}_f$ is fairly insensitive to q and a_f/r_0 and that

$$\hat{w}_f \leqslant \tfrac{1}{4}\,\frac{r_0 V_0^2}{c_0^2}\,. \tag{50}$$

Thus as a conservative and reasonable estimate of $\hat{w}_f$, for a wide range of values of q and a_f/r_0 we may take

$$\hat{w}_f = \tfrac{1}{4}\,\frac{r_0 V_0^2}{c_0^2} = \frac{r_0 I^2}{4mQ_0}\,. \tag{51}$$

It is of interest to compare this result with the corresponding results for an isotropic (unreinforced) rigid-plastic plate which were obtained by Wang [10] and Wang and Hopkins [11]. We denote the maximum final deflection using bending theory by $\hat{w}_B$. Then [10, 11]

$$\hat{w}_B = \frac{r_0^2 I^2}{8mM_0}, \qquad \textit{(simple support)},$$

$$\hat{w}_B = \frac{r_0^2 I^2}{14.28mM_0}, \qquad \textit{(clamped edge)},$$

where M_0 is the critical bending moment. For comparison we identify the initial shear yield stress k on planes z = constant with the shear yield stress of the unreinforced isotropic plate, with a Tresca yield condition. Then for a plate of thickness $2h$

$$Q_0 = 2kh, \qquad M_0 = 2kh^2,$$

and it follows that

$$\frac{\hat{w}_f}{\hat{w}_B} = \begin{cases} 2h/r_0 & \textit{(simple support)}, \\ 3.57h/r_0 & \textit{(clamped edge)}. \end{cases}$$

Hence the effect of ideal reinforcement is to reduce the deflection by a factor of the order of the aspect ratio of the plate. Identical results are obtained for a simply supported square plate.

2.3 Impact of a large plate by a mass

We now return to the case of a general yield function $F(Q_x, Q_y)$, and consider a certain special class of deformations. The presentation is based on [6]. We denote

$$F(x,y) = r, \tag{52}$$

where F is the same function as that which appears in the yield condition. Here r is a function of x and y with the dimension of a length, but is not, in general, a radial coordinate. We consider deformations of the form

$$w = h(r,t), \tag{53}$$

so that, at a given time, w is constant on each curve r = constant, and from (52) it follows that these curves of constant w in the x,y plane are similar to the yield curve in the Q_x, Q_y plane.

After some manipulation, which employs the convexity and homogeneity properties of F(x,y), we find

$$\bar{\gamma} = |\partial h/\partial r| . \tag{54}$$

Hence $\bar{\gamma}$, and therefore also W_p and Q_p, are constant on each curve F(x,y) = r. It is this property which gives particular interest to the class of deformations (53).

We now specialise further to deformations of the form

$$w = h(r,t) = \begin{cases} f(t) + g(r), & r < a(t), \\ 0 & r > a(t). \end{cases} \tag{55}$$

Then the region of the plate F(x,y) < a(t) moves as a rigid body with the remainder at rest. The curve F(x,y) = a(t) is taken as the curve C and v and γ_n are discontinuous across this curve. Since w is continuous at r = a(t), from (55) we have

$$f(t) + g\{a(t)\} = 0$$

and hence, by differentiating with respect to t and using (54), with suitable sign conventions, we obtain

$$v - \bar{\gamma}(a)\dot{a} = 0. \tag{56}$$

By using (56), the kinematic jump condition, the flow rule (43) and the homogeneity property of $F(Q_x, Q_y)$, it can then be shown that

$$c\frac{\partial r}{\partial n} = \dot{a}, \qquad Q_n = -\frac{Q_p}{\partial r/\partial n}, \tag{57}$$

where $\partial r/\partial n$ is the directional derivative of r in the direction $\underset{\sim}{n}$. Then the dynamic jump condition gives

$$Q_1 |\bar{\gamma}(a)|^n = m\dot{a}v. \tag{58}$$

Now (56) and (58) are two equations for $v(t)$, $\bar{\gamma}(a)$ and $a(t)$, which are the main unknown quantities. The third equation is the equation of motion of the interior of C and any mass M adhering to this region. After some manipulation this equation of motion reduces to

$$(M+mAa^2)\dot{v} = -2AaQ_p = -2Aa(Q_0+Q_1|\bar{\gamma}|^n), \tag{59}$$

where A is the area enclosed by $F(x,y) = 1$.

We now suppose that the mass M strikes the plate at the origin at time $t = 0$ and subsequently adheres to it. Then the governing equations are (56), (58) and (59) with the initial conditions

$$v = 0, \qquad a = 0, \qquad \text{when} \quad t = 0.$$

The solution is given in [6]. For brevity we record here only the solution for linear strain-hardening, which is

$$a = c_1 t, \qquad \text{where} \quad c_1^2 = Q_1/m,$$

$$v = \frac{AQ_0(a_f^2-a^2)}{c_1(M+mAa^2)},$$

$$\bar{\gamma}(a) = -\frac{\partial w(a)}{\partial r} = \frac{AQ_0(a_f^2-a^2)}{c_1^2(M+mAa^2)},$$

where

$$a_f^2 = \frac{c_1 MV_0}{AQ_0},$$

and a_f is the value of a when the plate comes to rest, and

$$\frac{mc_1^2 w}{Q_0} = \left(\frac{mA}{M}\right)^{\frac{1}{2}}\left(a_f^2 + \frac{M}{mA}\right)\left\{\tan^{-1}\left(\frac{mAa^2}{M}\right) - \tan^{-1}\left(\frac{mAr^2}{M}\right)\right\} - (a-r), \qquad \text{for} \quad r \leqslant a.$$

If the plate is of finite extent and M and V_0 are sufficiently large the discontinuity curve may propagate to the edge and be reflected back into the plate. Some cases in which this happens are discussed in [6].

For discussions of the limitations of the theory for the dynamics of ideal fibre-reinforced rigid-plastic beams and plates, and the approximations it involves, we refer to [5] and [6].

REFERENCES

[1] SPENCER, A.J.M., *J. Mech. Phys. Solids* 22 (1974) 147-159

[2] SPENCER, A.J.M., *Mechanics Res. Comms.* 3 (1976) 55-58

[3] SHAW, L. and SPENCER, A.J.M., *Int. J. Solids Struct.* 13 (1977) 823-831

[4] SHAW, L. and SPENCER, A.J.M., *Int. J. Solids Struct.* 13 (1977) 833-844

[5] SHAW, L. and SPENCER, A.J.M., *Int. J. Solids Struct.* 13 (1977) 845-854

[6] SHAW, L. and SPENCER, A.J.M., *Proc. R. Soc. London* A361 (1978) 43-64

[7] SPENCER, A.J.M., *Int. J. Eng. Sci.* 17 (1979) 35-47

[8] JONES, N., *J. Appl. Mech.* 98 (1976) 319-324

[9] LAUDIERO, F. and JONES, N., *J. Struct. Mech.* 5 (1977) 369-382

[10] WANG, A.J., *J. appl. Mech.* 22 (1955) 375-376

[11] WANG, A.J. and HOPKINS, H.G., *J. Mech. Phys. Solids* 3 (1954) 222-237

[12] LAUDIERO, F., *Università di Bologna, Facoltà di Ingegneria, Nota Tecnica* n.37 (1979)

X

NETWORK THEORY

A. C. PIPKIN
Division of Applied Mathematics
Brown University
Providence, Rhode Island 02912
U.S.A.

1. Introduction

The theory of networks of inextensible fibers was formulated by Rivlin [1] as a theory of materials like fish-nets, which have small resistance to distortion but high resistance to stretching. The theory is a continuum theory, in which individual cords or fibers are not recognized. The high resistance to stretching is idealized by treating the cords in the network as absolutely inextensible, and the low resistance to distortions is idealized by taking the shear modulus of the material to be zero.

Because Rivlin's theory uses the constraint that no fiber segment can change its length at all, even under compressive loading, solutions in this theory sometimes involves compressive stresses [2]. Rivlin recognized that this would happen, and warned that such solutions would not be physically meaningful. In the present lectures we outline an extended version of Rivlin's theory, in which fibers can grow shorter but not longer, and can carry tensile but not compressive loads.

After outlining Rivlin's analysis (as recast in vector form [2]) and his method of solution of pure traction boundary-value problems, we give some examples that illustrate why the theory needs to be extended. We then prove a uniqueness result that is not available for the theory in its original form, and close by giving some examples that illustrate the use of the uniqueness lemma.

The theory applies to closely-woven fabrics, in circumstances for which the resistance of the fabric to shearing deformations is of no importance. Adkins [3] formulated a theory of inextensible networks with elastic resistance to shearing deformations, and with a special strain-energy function Pipkin [4][5] has discussed the solution of traction boundary-value problems. In the brief outline presented here we do not discuss these elastic-deformation theories.

The examples that we discuss here are as simple as possible, to illustrate specific points of the theory. Rogers [6][7] has discussed some more complicated examples involving tears or holes in networks.

2. Networks of inextensible fibers.

We consider a sheet or net that, in its reference state, lies in some region of the X,Y plane. The network is composed of fibers that are parallel to the X and Y directions initially. The network is treated as a continuum, so that every line X = constant or Y = constant is regarded as a fiber.

In a deformation, the particle initially at $\underline{X} = X\underline{i} + Y\underline{j}$ goes to the place $\underline{x}(\underline{X}) = x(\underline{X})\underline{i} + y(\underline{X})\underline{j} + z(\underline{X})\underline{k}$, where $\underline{i}$, $\underline{j}$, and $\underline{k}$ are unit vectors parallel to the coordinate axes. Both families of fibers are governed by the same deformation function $\underline{x}(\underline{X})$ because the fibers are knotted together at the points where they cross, or, in the case of a woven fabric, held together by friction at these points.

The deformation maps a line element $\underline{i}dX$ onto $d\underline{x} = \underline{x},_X dX$, and an element $\underline{j}dY$ onto $d\underline{x} = \underline{x},_Y dY$. The vectors $\underline{x},_X$ and $\underline{x},_Y$ are tangential to the deformed fibers. For these vectors we use the notation

$$\underline{x},_X = \underline{a}, \quad \underline{x},_Y = \underline{b}. \qquad (2.1)$$

We wish to formulate a theory in which fibers cannot become longer in tension, but they can grow shorter by buckling or crumpling on the microscale, the scale of the actual distance between fibers in the physical network. Just as we do not take into account the actual wavy configuration of a thread in a woven fibric when we say that it lies parallel to the X-direction, we do not attempt to describe the details of fiber buckling in the function $\underline{x}(\underline{X})$ that describes the deformation.

Now, since the element $\underline{i}dX$ maps onto $\underline{a}dX$, its lengths before and after the deformation are dX and $|\underline{a}|dX$. We postulate that $|\underline{a}|$ cannot exceed unity. Similarity, $|\underline{b}|$ is no greater than unity:

$$\underline{a}\cdot\underline{a} \leq 1, \quad \underline{b}\cdot\underline{b} \leq 1. \tag{2.2}$$

When $|\underline{a}| = 1$ at some point, we say that the fiber is *extended* there. When $|\underline{a}| < 1$, the fiber is *slack*.

We limit our discussion to cases in which the network is loaded by forces around its edges only, with no external forces on the initially flat surface of the sheet. In such cases the equilibrium state of stress in the deformed network is conveniently described in terms of Rivlin's stress potential $\underline{F}(\underline{X})$.

Consider two curves running from $\underline{X}_A$ to $\underline{X}_B$. We mean material curves, and although we have in mind the curves in the deformed sheet, we describe them in terms of their configurations in the undeformed sheet. Let $\underline{F}_1$ be the force exerted by the material initially to the right of one of these curves on the material initially to its left, and let $\underline{F}_2$ be the similarly-defined force for the second curve. Then the resultant force on the region enclosed by the two curves is $\underline{F}_1-\underline{F}_2$ if the first curve was initially to the right of the second. In equilibrium this resultant must be zero, so the force exerted from right to left across a directed curve depends only on its endpoints. Let $\underline{F}(\underline{X})$ be the force that acts across any curve from $\underline{X}_o$ (any convenient origin) to $\underline{X}$. Then the force that acts across any curve from $\underline{X}_A$ to $\underline{X}_B$ is $\underline{F}(\underline{X}_B)-\underline{F}(\underline{X}_A)$.

The force acting across an infinitesimal element $d\underline{X} = \underline{j}dY$, which lies along $d\underline{x} = \underline{b}dY$ in the deformed state, is $\underline{F},_Y dY$. Thus $\underline{F},_Y$ is a stress vector, the force per unit initial length on a fiber initially in the Y-direction. Using $\underline{a}$ and $\underline{b}$ as base vectors, we can write

$$\underline{F},_Y = T_a\underline{a} + S\underline{b}. \tag{2.3}$$

Similarly, $-\underline{F},_X dX$ is the force exerted across an infinitesimal element $\underline{i}dX$, when it is in its deformed state $\underline{a}dX$, by the material initially above it on the material initially below it. We write $-\underline{F},_X$ in component form as

$$-\underline{F},_X = T_b\underline{b} + S'\underline{a}. \tag{2.4}$$

Any function $\underline{F}(\underline{X})$ describes a state of translational equilibrium. For rotational equilibrium, it is necessary and sufficient [2] for S and S' in (2.3) and (2.4) to be equal.

We now postulate that the network transmits forces only by means of tensions in the fibers. This means that $S = S' = 0$, so that

$$\underline{F},_Y = T_a\underline{a} \quad \text{and} \quad -\underline{F},_X = T_b\underline{b}\,. \tag{2.5}$$

For, the force on a fiber X = constant must be in the direction $\underline{a}$ of the fibers that cross it.

We also postulate that the tensions T_a and T_b cannot be negative:

$$T_a \geqq 0, \quad T_b \geqq 0. \tag{2.6}$$

When T_a is positive at some point, we say that the fiber is *tense* there. When $T_a = 0$, is it *relaxed*.

Finally, we postulate that a slack fiber must be relaxed:

$$\begin{aligned} |\underline{a}| < 1 \quad &\text{implies} \quad T_a = 0, \\ |\underline{b}| < 1 \quad &\text{imples} \quad T_b = 0. \end{aligned} \tag{2.7}$$

In the context of (2.2) and (2.6), this is equivalent to the statement that a tense fiber must be extended:

$$\begin{aligned} T_a > 0 \quad &\text{implies} \quad |\underline{a}| = 1, \\ T_b > 0 \quad &\text{implies} \quad |\underline{b}| = 1. \end{aligned} \tag{2.8}$$

An extended fiber may be either tense or relaxed, and a relaxed fiber may be either extended or slack. The only impossible combination is a tense, slack fiber.

The tensions T_a and T_b are reactions to the one-sided constraints (2.2). They are not specified by constitutive equations, but take whatever values equilibrium requires. In particular, as in other theories involving inextensible fibers, a single fiber may carry a finite force, giving a Dirac delta singularity in the corresponding tension T_a or T_b. In terms of the stress potential $\underline{F}$, such singularities are represented by finite discontinuities [2].

3. Boundary-value problems.

In a typical boundary-value problem the position $\underline{x}(\underline{X})$ is required to take specified values $\underline{x}_0(\underline{X})$ on a part C_p of the boundary C of the sheet, and tractions are specified on the remaining part, C_t. Tractions are specified most conveniently by giving the values of the stress potential $\underline{F}$ on C_t. Then the increment $d\underline{F}$ along a boundary arc $d\underline{X}$ is the external force on that arc. Thus, we consider problems with boundary conditions of the type

$$\underline{x}(\underline{X}) = \underline{x}_0(\underline{X}) \quad \text{on} \quad C_p, \quad \underline{F}(\underline{X}) = \underline{F}_0(\underline{X}) \quad \text{on} \quad C_t. \tag{3.1}$$

A problem specified in this way will fail to have a solution if there is no kinematically admissible deformation (satisfying the constraints (2.2)) with the boundary values $\underline{x} = \underline{x}_0$. In particular, if part of the boundary lies along a fiber, boundary values of $\underline{x}(\underline{X})$ along that fiber must not violate (2.2).

Problems often have infinitely many solutions, but the non-uniqueness is of an understandable, expected type. We discuss some examples of non-uniqueness later.

Four kinds of regions may appear in solutions. With some overlap, they are

a) *Slack* regions, where $|\underline{a}| < 1$ and $|\underline{b}| < 1$;

b) *Half-slack* regions, where the fibers in one family are extended and the fibers in the other family are slack;

c) *Collapsed* regions, in which $\underline{a}$ and $\underline{b}$ are parallel at every point;

d) *Fully extended* regions, where $\underline{a}$ and $\underline{b}$ are non-parallel unit vectors.

In a slack region the tensions are unique, namely zero, but the deformation is highly non-unique. A proper uniqueness theorem might state the uniqueness of the region, but not uniqueness of the deformation in it.

In a half-slack region in which the extended fibers are also tense, they are straight, and the tension in each fiber is constant along it.

In a collapsed region with tense fibers, all of the fibers lie along the same straight line in the deformed state [2]. The distribution of the total load among these fibers is highly non-unique. An example is given in Section 8.

Rivlin's theory [1] is, in effect, the theory of plane deformations of fully extended regions. Some of Rivlin's results are outlined in the following Section.

4. Plane, fully extended regions

Rivlin's [1] theory is restricted to plane deformations and uses the assumptions that $\underline{a}$ and $\underline{b}$ are non-parallel unit vectors. In a plane deformation the deformed sheet lies in the plane $\underline{k}\cdot\underline{x} = 0$, say, so $\underline{k}\cdot\underline{a} = \underline{k}\cdot\underline{b} = 0$. In any deformation, plane or not,

$$\underline{a},_Y = \underline{x},_{XY} = \underline{b},_X, \tag{4.1}$$

and in any deformation with $\underline{a}\cdot\underline{a} = \underline{b}\cdot\underline{b} = 1$,

$$\underline{a}\cdot\underline{a},_Y = \underline{b}\cdot\underline{b},_X = 0. \tag{4.2}$$

Then $\underline{a},_Y$ is orthogonal to $\underline{a}$, $\underline{b}$, and $\underline{k}$, so if $\underline{a}$ and $\underline{b}$ are not parallel, $\underline{a},_Y$ is zero. Similarly, $\underline{b},_X = \underline{0}$. Thus

$$\underline{a} = \underline{a}(X) \quad \text{and} \quad b = \underline{b}(Y). \tag{4.3}$$

$$\underline{x} = \underline{f}(X) + \underline{g}(Y). \tag{4.4}$$

This means that the fibers in either family lie along congruent curves.

From (2.5),

$$\underline{k}\text{x}\underline{a}(X)\cdot\underline{F}_{,Y} = 0 \quad \text{and} \quad \underline{k}\text{x}\underline{b}(Y)\cdot\underline{F}_{,X} = 0. \tag{4.5}$$

Integration gives

$$\underline{k}\text{x}\underline{a}\cdot\underline{F} = M(X) \quad \text{and} \quad \underline{k}\text{x}\underline{b}\cdot\underline{F} = N(Y). \tag{4.6}$$

Then since

$$\underline{F} = J^{-1}[\underline{b}(\underline{k}\text{x}\underline{a}\cdot\underline{F}) - \underline{a}(\underline{k}\text{x}\underline{b}\cdot\underline{F})], \tag{4.7}$$

where

$$J = \underline{k}\cdot\underline{a}\text{x}\underline{b}, \tag{4.8}$$

it follows that

$$\underline{F} = J^{-1}[M(X)\underline{b}(Y) - N(Y)\underline{a}(X)]. \tag{4.9}$$

The solution of dead-loading traction boundary value problems is particularly simple. Suppose that tractions are prescribed on C by specifying the force per unit initial length on every boundary arc. Integrating these tractions with respect to arc length along the undeformed boundary gives the boundary values of $\underline{F}$. Of course, for a plane problem, these boundary values must satisfy $\underline{k}\cdot\underline{F} = 0$.

Let $\Delta\underline{F}(X)$ be the total traction on the part of the boundary initially to the right of the fiber X, and let $-\Delta\underline{F}(Y)$ be the force on the part of the boundary initially above the fiber Y. These can be expressed in terms of the boundary values of $\underline{F}$ by

$$\Delta\underline{F}(X) = \underline{F}(X,Y_+) - \underline{F}(X,Y_-) \tag{4.10}$$

and

$$\Delta\underline{F}(Y) = \underline{F}(X_+,Y) - \underline{F}(X_-,Y). \tag{4.11}$$

Here Y_+ and Y_- are the places where the fiber X crosses the boundary, and X_+ and X_- are the places where the fiber Y crosses the boundary.

Now, the force $\Delta\underline{F}(X)$ is transmitted across the fiber X by tensions parallel to $\underline{a}(X)$, so

$$\underline{a}(X) = \Delta\underline{F}(X)/|\Delta\underline{F}(X)|. \qquad (4.12)$$

Similarly,

$$\underline{b}(Y) = -\Delta\underline{F}(Y)/|\Delta\underline{F}(Y)|. \qquad (4.13)$$

When the two fields $\underline{a}$ and $\underline{b}$ have been found, the two functions M and N in the expression (4.9) for $\underline{F}$ are found by using the boundary values of $\underline{F}$ again in (4.6). Evidently M(X) can be determined from the boundary values at either (X,Y_+) or (X,Y_-). Consistency of these two evaluations requires that

$$\underline{k}\times\underline{a}(X)\cdot\Delta\underline{F}(X) = 0, \qquad (4.14)$$

but this is satisfied because $\underline{a}$ is parallel to $\Delta\underline{F}(X)$.

The assumptions used in the analysis must now be checked. The first opportunity for a contradiction arises when J is calculated from (4.8). If J = 0 throughout some region, (4.9) is nonsense. If $J \neq 0$, the tensions T_a and T_b must be calculated from (4.9) by using (2.5). If either tension is negative somewhere, the solution is invalid. Finally, at boundaries along fibers, the value of $\underline{F}$ given by (4.9) is likely to differ from the prescribed boundary values by a term parallel to the boundary [2]. This term represents a finite force carried by the boundary fiber, and it must correspond to a tension rather than a compressive load. If it is a compression, again the solution is invalid.

If the solution obtained by the resultant-force method is found to be invalid for any of these reasons, then either the network is fully extended but contains a fold, or it is not fully extended everywhere. We give examples of these cases in the following Sections.

5. Folds in fully extended regions [2].

The resultant-force method of determining $\underline{a}$ and $\underline{b}$ does not work if the deformed sheet contains a fold across which $\underline{a}$ and $\underline{b}$ are discontinuous. At a fold that cuts across both families of fibers, the tangential components of $\underline{a}$ and $\underline{b}$ are continuous but the normal components change sign. The forms of the functions $\underline{a}(X)$ and $\underline{b}(Y)$ change from $\underline{a}_+(X)$ and $\underline{b}_+(Y)$ on one side of the fold to $\underline{a}_-(X)$ and $\underline{b}_-(Y)$ on the other side. Since $\underline{a}$ is not constant over the whole length of the fiber X, the direction of the resultant force $\Delta\underline{F}(X)$ does not give the direction of $\underline{a}$.

As compensation for the complication introduced by the suspected presence of a fold, the stress field near a fold is particularly simple: the tensions T_a and T_b are zero in any rectangle bounded by fibers with the fold connecting diagonally opposite corners. (By "rectangle", we mean the configuration of the region in the reference state.) For, first, fibers can have no tension at the place where they cross the fold, because in the deformed state these tensions would all pull toward the same side and the fold locus would not be in equilibrium. Then second, for any triangular region bounded by the fold and two fibers X and Y that cross it, the forces on the region are in the directions $\underline{a}(X)$ and $\underline{b}(Y)$, and there is no third force to equilibrate these, so they must be zero.

(An *envelope* of fibers is a fold at which the fibers are tangential to the fold and thus to one another. Fiber tensions need not vanish at an envelope. Envelopes are curves along which $J = 0$ in (4.9).)

As an example, consider a square sheet with point forces at its corners (Figure 1a). The forces at the corners (0,0),(L,0),(L,L), and (0,L) are respectively $-2F\underline{j}$, $F(\underline{j}-\underline{i})$, $2F\underline{i}$, and $F(\underline{j}-\underline{i})$. The resultant-force method yields the result that both $\underline{a}$ and $\underline{b}$ are parallel to $\underline{i}+\underline{j}$. This gives $J = 0$ and shows that the analysis in Section 4 is not applicable.

Inspection of the boundary loads suggests that the sheet will fold along the line $X = Y$ (Figure 1b). This yields a consistent solution in which the whole interior of the sheet is relaxed (tension-free) and the boundary loads are supported solely by finite forces in the boundary fibers.

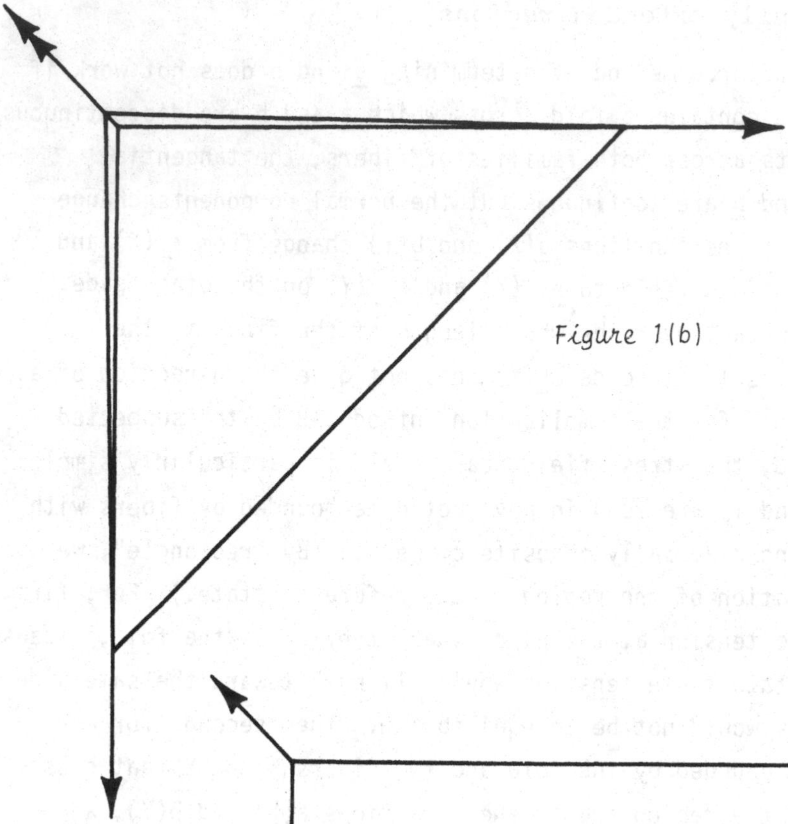

Figure 1(b)

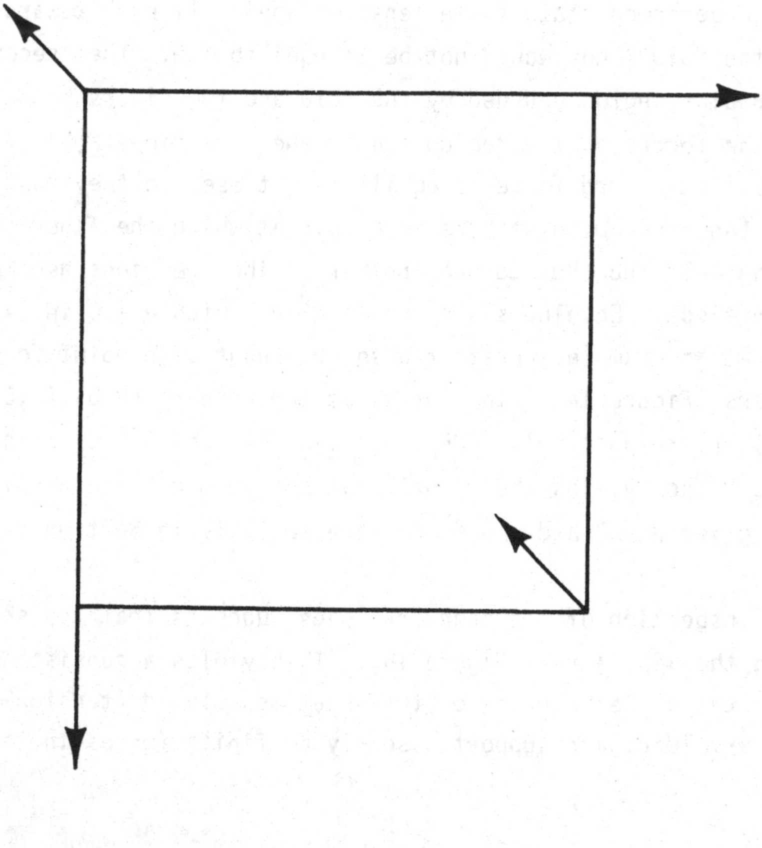

Figure 1(a)

6. Negative tensions, suggesting a slack region.

In some cases the hypothesis that the sheet is fully extended will lead to a contradiction even when the possible presence of folds is taken into account.

For example, consider a rectangular sheet loaded with uniform normal tensions over most of its boundary, but with a small compressive loading on a small part of the boundary (Figure 2). The applied tractions are $\pm\underline{j}T_2$ on the top and bottom edges $Y = \pm H$, and are $\pm\underline{i}T_1$ on the ends $X = \pm L$, except in the region $|Y| < h$. In the latter regions, on $X = L$ the traction is $-P\underline{i}$, and on $X = -L$ it is $P\underline{i}$.

If hP is smaller than $(H-h)T_1$, the resultant-force method gives $\underline{a} = \underline{i}$, and it gives $\underline{b} = \underline{j}$ in any event. Then the sheet is undeformed, and the tensions in it are easily seen to be $T_b = T_2$ and $T_a = T_1$ or $-P$, depending on the value of Y. The result that $T_a = -P$ for $|Y| \le h$ contradicts the hypothesis that the deformed sheet is a fully extended region with no folds.

We might next try the hypothesis that the sheet is fully extended but has a fold somewhere. This is not a natural hypothesis, but how can we find out whether or not there is a solution of that type, without examining every conceivable configuration of folds?

It is possible to construct a solution with slack regions near the points $(\pm L,0)$ (Figure 3). This answers our worries about solutions with folds, because there is a uniqueness lemma which, applied to this case, implies that the solution with slack regions is essentially unique. Of course, the deformation in the slack region is not unique, so the uniqueness lemma does not say the same sort of thing that uniqueness theorems usually say.

7. Uniqueness [8].

We will show that if a given problem has a solution, that solution minimizes the energy of the boundary loads. We will then use the minimum-energy property to show that if a fiber is tense at some point, in some solution, then it is fully extended at that point in every

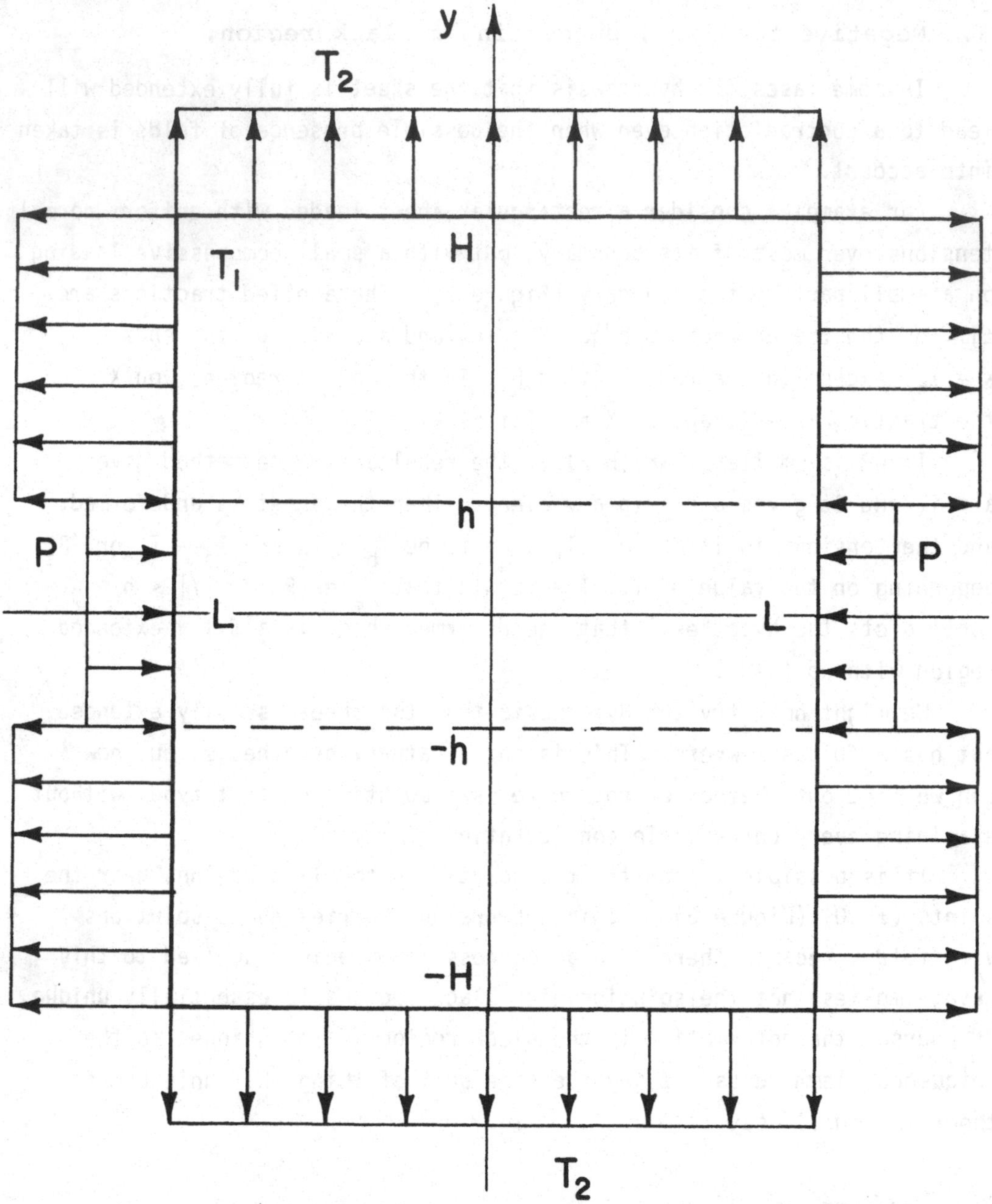

Figure 2

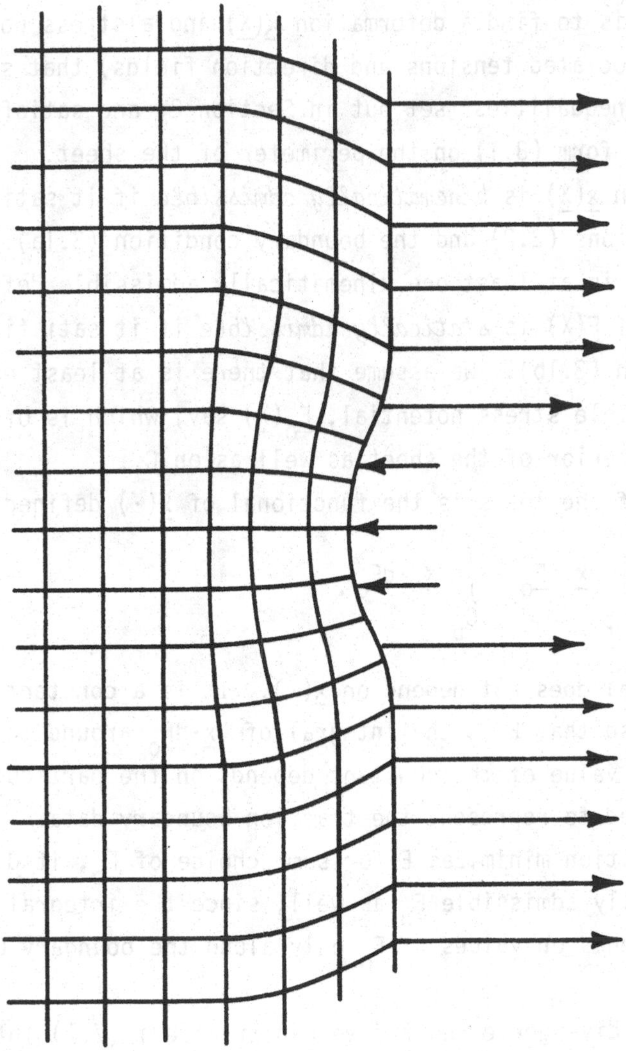

Figure 3

solution, and its direction at that point is uniquely determined. It follows that if a fiber is slack at some point, in some solution, the tension in it is zero at that point in every solution.

The problem is to find a deformation $\underline{x}(\underline{X})$ and a stress potential $\underline{F}(\underline{X})$ (and the associated tensions and direction fields) that satisfy the equations and inequalities set out in Section 2, and satisfy boundary conditions of the form (3.1) on the perimeter of the sheet.

A deformation $\underline{x}(\underline{X})$ is *kinematically admissible* if it satisfies the constraint conditions (2.2) and the boundary condition (3.1a). We assume that there is at least one kinematically admissible deformation. A stress potential $\underline{F}(\underline{X})$ is *statically admissible* if it satisfies the boundary condition (3.1b). We assume that there is at least one statically admissible stress potential, $\underline{F}_o(\underline{X})$ say, which is of course defined on the interior of the sheet as well as on C_t.

The energy of the loads is the functional of $\underline{x}(\cdot)$ defined by

$$E[\underline{x}(\cdot)] = -\int_{C_t} \underline{x}\cdot d\underline{F}_o - \int_{C_p} \underline{x}_o\cdot d\underline{F}_o \ . \qquad (7.1)$$

The second integral does not depend on $\underline{x}(\cdot)$. It is a constant thrown in for convenience, so that E is the integral of $-\underline{x}\cdot d\underline{F}_o$ around the whole perimeter C. The value of the constant depends on the particular stress function $\underline{F}_o(\underline{X})$ used to represent the traction boundary data. However, if a given deformation minimizes E for some choice of $\underline{F}_o$, it does so for any other statically admissible $\underline{F}_o$ as well, since the integral involving $\underline{x}(\cdot)$ in (7.1) depends on values of $\underline{F}_o$ only along the boundary C_t where $\underline{F}_o$ is prescribed.

By using the divergence theorem we can transform (7.1) into an integral over the region occupied by the sheet in its reference state:

$$E[\underline{x}(\cdot)] = -\oint_C \underline{x}\cdot(\underline{F}_{o,X}dX + \underline{F}_{o,Y}dY)$$

$$= \iint[(\underline{x}\cdot\underline{F}_{o,X})_{,Y} - (\underline{x}\cdot\underline{F}_{o,Y})_{,X}]dXdY. \qquad (7.2)$$

Now, let T_a^o, T_b^o, $\underline{a}_o$, and $\underline{b}_o$ be the artifical tensions and direction fields derived from $\underline{F}_o$ by using (2.5). That is, T_a^o is the magnitude of $\underline{F}_{o,Y}$, and wherever this magnitude is not zero, $\underline{a}_o$ is a unit vector in the direction of $\underline{F}_{o,Y}$. Also, let $\underline{a}$ and $\underline{b}$ be given in terms of $\underline{x}$ by (2.1) as usual. Then (7.2) becomes

$$E[\underline{x}(\cdot)] = -\iint [T_a^o \underline{a}_o \cdot \underline{a} + T_b^o \underline{b}_o \cdot \underline{b}] dXdY. \tag{7.3}$$

Because the tensions are non-negative and none of the vectors exceed unity in magnitude, it is apparent that

$$E[\underline{x}(\cdot)] \geqq -\iint (T_a^o + T_b^o) dXdY, \tag{7.4}$$

with equality only if $\underline{a} = \underline{a}_o$ wherever $T_a^o \neq 0$ and $\underline{b} = \underline{b}_o$ wherever $T_b^o \neq 0$. Since $\underline{a}_o$ and $\underline{b}_o$ were derived from an arbitrary admissible stress function rather than a deformation, there is generally no deformation $\underline{x}_o(\underline{X})$ from which they can be derived, so the choice of $\underline{a}$ and $\underline{b}$ that gives equality in (7.4) does not correspond to an admissible deformation.

However, suppose that the problem (specified by the equations and inequalities in Section 2 and the boundary conditions (3.1)) has a solution. Let this solution be denoted by $\underline{x}_o(\underline{X})$, $\underline{F}_o(\underline{X})$ on the interior as well as on the boundary. Then using this $\underline{F}_o$ in the definition of E, the vectors $\underline{a}_o$ and $\underline{b}_o$ derived from it are also derivable from the deformation $\underline{x}_o(\underline{X})$, so the choice $\underline{a} = \underline{a}_o$ and $\underline{b} = \underline{b}_o$ is admissible in (7.3). Thus, the deformation $\underline{x}_o(\underline{X})$ minimizes E.

We emphasized earlier that if a given deformation minimizes E, it does so independently of the particular statically admissible stress function $\underline{F}_o$ used in defining E. Consequently, every solution $\underline{x}(\underline{X})$ minimizes E. But this means that if T_a^o, T_b^o, $\underline{a}_o$, and $\underline{b}_o$ are the tensions and fiber directions for some solution, and $\underline{a}$ and $\underline{b}$ are the fiber directions for some possibly different solution, then $\underline{a} = \underline{a}_o$ wherever $T_a^o > 0$ and $\underline{b} = \underline{b}_o$ wherever $T_b^o > 0$. Thus, $\underline{a}$ is uniquely determined at any point where T_a is positive in some solution, and $\underline{b}$ is uniquely determined at any point where T_b is positive in some solution.

8. Examples.

The set of fibers or fiber segments that are tense in *some* solution of a problem, even if not tense in all solutions, is the *load-carrying part* of the network for that problem. The uniqueness lemma proved in Section 7 says roughly that the deformation of the load-carrying part is unique. Stronger uniqueness results are true for more restricted classes of problems than the class considered in Section 7, but for that general class, there are good reasons why the uniqueness lemma says no more than it does.

For example, it says nothing about uniqueness of tensions. To see that tensions need not be unique, suppose that a sheet is clamped along its undeformed boundary, so that $\underline{x} = \underline{X}$ is one solution. In this solution the fiber tensions T_a and T_b are arbitrary non-negative functions of Y and X, respectively.

In this problem, is $\underline{x} = \underline{X}$ the only solution? It is geometrically obvious that it is in fact the only kinematically admissible deformation, but proving this analytically looks difficult. However, we can prove it immediately by using the uniqueness lemma. For each fiber X there is a solution in which the tension in that fiber is positive and $\underline{b} = \underline{j}$, so $\underline{b} = \underline{j}$ in every solution. Similarly, $\underline{a} = \underline{i}$ in every solution because there is a solution with $\underline{a} = \underline{i}$ and positive tensions. So $\underline{x} = \underline{X} + \underline{c}$, and the constant $\underline{c}$ must be zero to satisfy the boundary conditions.

Non-unique tensions are naturally expected in problems with no boundary tractions specified, but the tensions can also be non-unique even in pure traction boundary-value problems. This happens when a *collapsed* region occurs in the solution. For example, consider a square sheet with boundaries along $X = \pm L$ and $Y = \pm L$, loaded by point forces $\underline{f}$ and $-\underline{f}$ at the corners (L,L) and (-L,-L), respectively, and with the rest of the boundary traction-free. The resultant-force method suggests that $\underline{a} = \underline{b} = \underline{f}/|\underline{f}|$. Let $\underline{F} = \underline{f}\phi(X,Y)$, where ϕ is any function that increases monotonically from zero to one as Y increases from -L to L, and decreases monotonically from one to zero as X increases from -L to L. Then $T_a = |\underline{f}|\phi,_Y$ and $T_b = -|\underline{f}|\phi,_X$. All conditions are satisfied, with no further restriction on ϕ. For every point of every fiber there is a solution in which that fiber is tense, so $\underline{a} = \underline{b} = \underline{f}/|\underline{f}|$ is the unique solution for the fiber directions. Then $\underline{x} = (X+Y)\underline{a} + \underline{c}$. The whole sheet collapses onto a straight line in the direction of $\underline{f}$. The total force $\underline{f}$ is distributed among the fibers in an indeterminate way.

The fiber direction may be unique even in fibers that do not belong to the load-carrying part of the network. For example, consider a sheet loaded as in Figure 2, with P = 0. The uniqueness lemma, applied to the obvious solution, shows that $\underline{b} = \underline{j}$ and that $\underline{a} = \underline{i}$ for $|Y| > h$, in *every* solution. Then $\underline{x} = \underline{X} + \underline{c}$ is the most general solution. Consequently, $\underline{a} = \underline{i}$ for $|Y| < h$ as well, although these fibers carry no tension in any solution.

The uniqueness lemma can be used to prove uniqueness of tensions when those tensions are zero. For example, suppose that tractions are specified as zero on the whole boundary of a sheet. Then $\underline{x} = \underline{0}$ is a solution in which the whole sheet is slack. This proves that $T_a = T_b = 0$ in every solution. Every deformation satisfying (2.2) is admissible, so the deformation is highly non-unique.

In all of the previous examples, the results are unsurprising. The main use of the uniqueness lemma is to make it trivial to prove unsurprising results.

ACKNOWLEDGMENT

This paper was prepared with the support of a grant MCS 79-03392 from the National Science Foundation. We gratefully acknowledge this support.

REFERENCES

[1] RIVLIN, R.S. Plane strain of a net formed by inextensible cords. Arch. Rat. Mech. Anal. 4, (1955) 951-74.

[2] PIPKIN, A.C. Some developments in the theory of inextensible networks. Q.Appl. Math. 38, (1980) 343-55.

[3] ADKINS, J.E. Finite plane deformation of thin elastic sheets reinforced with inextensible cords. Phil. Trans. Roy. Soc. London 249, (1956) 125-50.

[4] PIPKIN, A.C. Plane traction problems for inextensible networks. QJMAM 34 (1981) 415-29.

[5] PIPKIN, A.C. Finite plane stress of stiff fiber-reinforced sheets. JIMA 27 (1981) 195-209.

[6] ROGERS, T.G. and PIPKIN, A.C. Holes in inextensible networks. QJMAM 33 (1980) 447-62.

[7] ROGERS, T.G. Crack extension and energy release rates in finitely deformed sheets reinforced with inextensible fibers. Dept. of Theoretical Mechanics, Univ. of Nottingham, 1981.

[8] PIPKIN, A.C. Inextensible networks with slack. Q. Appl. Math. 40 (1982) 63-71.

Printed in the United States
By Bookmasters